The Secret Languages
of Animals

The Secret Languages of Animals

REVISED AND UPDATED

VINSON BROWN

Illustrations by Judy Wilmot

Prentice Hall Press • New York

To my granddaughter, Fawn Brown,
because of her great interest in animal life

Copyright © 1955, 1987 by Vinson Brown
All rights reserved, including the right of reproduction
in whole or in part in any form.

Published in 1987 by Prentice Hall Press
A Division of Simon & Schuster, Inc.
Gulf+Western Building
One Gulf+Western Plaza
New York, NY 10023

Originally published as *How to Understand Animal Talk* by Little, Brown & Co.

PRENTICE HALL PRESS is a trademark of Simon & Schuster, Inc.

Library of Congress Cataloging-in-Publication Data

Brown, Vinson, 1912-
The secret languages of animals.

Rev. and updated ed. of: How to understand animal talk.
Bibliography: p.
Includes index.
1. Animal communication. 2. Human-animal
communication. I. Brown, Vinson, 1912-
How to understand animal talk. II. Title.
QL776.B76 1986   591.59   86-25164
ISBN 0-13-798026-4

Manufactured in the United States of America

Designed by Irving Perkins Associates

10 9 8 7 6 5 4 3 2 1

First Prentice Hall Press Edition

Contents

	Preface	vii
1	Entering a Secret World	1
2	A Scientific Approach	7
3	Domestic Animals	25
4	The Meat Eaters	56
5	Vegetarian Mammals	100
6	Wild Birds	131
7	Reptiles, Amphibians, and Fish	159
8	Insects and Their Relatives	169
9	Finding the Answers to the Unknown	190
	Index	195

Preface

In this book, a revision of *How to Understand Animal Talk,* I have tried to explore more deeply the fascinating subject of animal communication, or language.

Everyone knows that animals, with rare exceptions, do not use speech as humans do. They show emotions by sounds, movements, smells, and even tastes, but not in spoken ideas in the sense that men do. A dog could not say to another dog: "I saw you downtown yesterday with your master." Rather, by wagging his tail, he expresses the emotion: "I am glad to see you now!"

Animal talk is almost entirely emotional, as when we grunt with pain or move our facial muscles into a frown or smile. It is largely talk about the present, without past or future. As an expression of emotion, on the other hand, it is usually far more complex and subtle than the rather rare emotional talk of men, and for this reason large parts of it often remain mysterious to us.

There has been considerable research done on animal communication in the twelve years since I first wrote about it, and in this edition I have tried to refine the categories of animal communication and how these categories relate to one another.

Animal behavior is enormously complex and cannot possibly be described adequately in a single book, but I hope you will come away from your reading with sufficient knowledge and genuine interest to further study this fascinating subject. Most people put mental barriers between themselves and the lesser creatures of this world, but we can break through these walls to understand more fully a wonderful new world that needs our protection and understanding.

1
Entering a Secret World

In the beginning, say some of the old legends, all the animals could talk and think just like men. Then somehow, through the foolishness of one animal or another, a catastrophe came, and they lost this ability. Other legends say that once men knew the language of animals and could converse easily with them, but some supernatural force punished men for their foolishness by taking this magic ability away from them. Primitive tribes even to this day send out their young men to make contact with certain animals, to learn their languages and receive magic power from them.

Once in the jungles of Costa Rica near Punta Burica, I watched an Indian named Pi-tin-tin call wild animals and birds out of the forest. So perfectly like the calls and whistles of the wild creatures were the sounds he made that many creatures came to within a few yards of us, calling in answer.

If we were like Pi-tin-tin or Mowgli, the boy of Kipling's *Jungle Book* who was raised with the wolves of India, we would have the feel of the wilderness in our bones, and it would be fairly easy for us to pick up the languages of animals. But most of us have lived much of our lives between four walls and have had our senses blunted by too many civilized smells and noises. It will not be easy for us to enter this secret world of animal talk, but it will be fun trying.

To understand animals successfully, we must feel close enough to them to catch their feelings and thoughts by a kind of mental telephone system. When we are close enough to animals in this way, they feel friendly toward us and often teach us many things that they would keep completely hidden from an unsympathetic person.

Furthermore, we must learn to listen. Persons who can go into the woods or fields alone, and who can be quiet and listen, draw from the world of nature an inner strength and wisdom that will serve them all their lives. Be quiet, very quiet, and listen intently. Who knows what strange secrets you may discover?

Learn to watch carefully. The eyes of the ordinary city dweller are not focused to the movements of animals, wild or domesticated. You must train yourself to sit or walk quietly and keep your eyes alert to every movement around you. Most animals' movements, even their very tiny movements and the expressions in their eyes, have meanings.

Try to revive that sense that man has almost lost, the sense of smell. All sorts of strange little tales come down the wind to your pet dog, but to you they mean nothing. You can help yourself a little by wetting your nose with your finger and sniffing deeply. Keep trying, and you will gradually learn to recognize some of the stronger scents of the woods and fields.

One of the primary purposes of this book is to help you put sounds, smells, and animal signs into separate types so you can more quickly understand their meanings. Animals make different sounds at mating time than at other times of the year. Herd and pack sounds are different and have different meanings from those of individual animal sounds. Sounds of rage and fear or of warning must be separated from each other and from other sounds. Animal movements and smells fit into different types, too. As you learn about the languages of different animals in this book, notice the many different types and meanings of animal talk and learn to recognize these in the wild.

One day a pet cat came to a friend of mine and began to rub against his legs, meowing in a peculiar way. When he paid no attention to her, she nipped his leg very gently and mewed more loudly. My friend was naturally kind to animals, so he said, "What is the matter, Kitty?" He patted her and listened to her. Her meow rose in crescendo, and rising on her hind legs, she reached up almost imploringly toward him.

He decided that she was soon going to have her kittens. So he found a wooden box and prepared a nest in it of old clothes. He set the box on the back porch, and the cat immediately jumped into it.

A cat calls for food.

She turned around about three times to get the feel of it, and then jumped out with a loud purr and began to rub against his legs. As plainly as possible, she was saying "Thank you!" Sure enough, that night she had nine kittens!

Recently on our ranch I heard a mother killdeer calling, down by the fence where the overflow from our water tank keeps the grass and weeds green all through the year. This beautiful white-and-black-marked shore bird was crying "Kee-dee-dee!" over and over. There was a soft, gentle, encouraging inflection in the sound that roused my curiosity, so I went down very quietly to see what was going on.

Some tiny baby killdeers, scarcely more than fluffs of black and white, were being urged by their mother to hunt insect food with her among the weeds. But the instant the mother became aware of my presence, the tone of her voice changed. The "Kee-dee-dee!" took on a high, frightened sound, and the babies scattered among the grass, seeking the best shelter they could find. Two of them ran in my direction, and I picked them up.

Again the mother's voice changed. This time the "Kee-dee-dee!" expressed pain and helplessness. The mother was across the fence from me, dragging her wing on the ground and acting as if severely wounded. When I paid no attention to her, she flew closer to me and again dropped to the ground, crying piteously and dragging her wing.

I approached the crying mother and set the tiny killdeers free. Away they rushed through the grass, under the fence and away from me. When the mother was sure they were safe, she stopped her pitiful cries and her wounded act. Into the air she sprang winging swiftly to where the babies were. As I walked back toward our house, I heard her triumphant cry: "Kee-dee-dee! kee-dee-dee!"—still the same apparent sounds, but what a difference in the inflection!

We can translate her messages into human words somewhat as follows: (*First*) "Come, children, come; here is something good to eat!" (*Second*) "Look out! An enemy is coming; scatter and hide!" (*Third*) "Look over here, Mr. Man, see how wounded I am. If you would chase me and leave my babies alone, you could catch me for sure!" (*Fourth*) "How glad I am! How glad I am! My babies are safe! I fooled the man!"

The killdeer language by inflection is a language of emotions, emotions that are comparable to ours but that we can express only crudely in words. The meaning to the killdeer of these cries is far richer than it is to us, so that always in our attempts to understand,

A killdeer fakes a broken wing to protect her young.

we will fall short of the full and true meaning. Nevertheless, through sympathy and love and careful observation, we can come to know much about the emotional languages of wild creatures.

This book gives information about the languages of representative common animals that live in the United States and Canada. Naturally, you will occasionally hear and see animals whose languages are not described in this book. But if you study the book carefully, you will be much better prepared to understand even such unknown languages. If you read about the language of one kind of mouse, you will have knowledge that will help you understand similar kinds.

The main thing to remember is that there are usually differences between the languages even of animals that are closely related. One kind of mouse may be much more aggressive and unafraid than another kind. Its language will show that aggressiveness by being louder or bolder or more insistent. If you study the life of the animal closely, you will sense the difference and be able to interpret it correctly. By listening carefully, you can detect differences in tone. The animal's action will probably explain what those tones mean. For example, while it runs away, the noise it makes obviously means that it is afraid. Try to put yourself inside the skin of the animal so you can understand what emotions it is expressing.

Since the languages of most animals are at least in part instinctive, it is a good idea for us to know just what "instinctive" means. An instinctive language is an unlearned language. Thus, a puppy that is raised completely away from other dogs will bark, growl, whine, wag its tail, and wriggle with delight without being taught to do any of these things. However, a dog that has experience with other dogs learns how to use all these signals better and more expressively than the dog who is raised alone.

Some animal languages go beyond instinct. A few highly intelligent animals, such as the wolf or the crow, learn a good deal of their language by copying adults, as human children do. Thus the young crow instinctively knows the common "Caw!" cry of the crow, but it learns from its parents and from other adults all sorts of inflections of this cry that have different shades of meaning. So part of crow language is learned as well as instinctive.

Animals communicate with calls, songs, and colors (including colors that can be changed voluntarily, as when an antelope flashes its white rump patch in the sunlight by moving the hairs). Animals also express themselves in movements that are sometimes so small as to be noticed only by very sharp and watchful human eyes, but that have rich meanings between animals of the same kind or species. Animals also use smells to leave and send signals. It is necessary to keep all these means of communication in mind as you watch any animal.

Notice how the kind of weather and the time of day, night, or year affect animals and their languages. Squirrels and chipmunks stop chattering and hole up for a nap during bad weather, for example. A cold day may cause a bear to go to sleep or into hibernation, but it will make a marten or weasel more active than ever. Some birds, such as the whippoorwills and owls, are heard mainly at night and are active then. Others, such as hummingbirds and sparrows, are active only during daylight.

During their mating seasons, animals behave very differently than they do at other times. Bull elk and buck deer become short-tempered and dangerous in the fall when they seek mates. California woodpeckers fill the days of spring with constant noise as they seek mates, but become very quiet later on to hide their movements when feeding young.

Territory is very important to some animals, and it influences their use of language. In the spring many birds stake out special territories by using their songs to warn other birds of the same kind to stay away. Many lizards also claim territory and use different ways to drive rivals away.

Remember that each living place (or habitat) has an influence on the animals that live there. You will hear birds scream on the seacoast, where they must be heard above the roar of the waves. You will see animals of the Great Plains depending much more on eyesight and distant signals than do animals of the deep woods, which rely more on their ears and their noses for awareness of the approach of friends or enemies. Animals living under water depend heavily on observing each other's movements, especially as they flash in the sunlight. If you wish to understand animal languages, watch for all these things.

2
A Scientific Approach

It is impossible to write a book of this size about animal languages and cover more than a small part of the whole picture. Even the scientists who study animal behavior, the ethologists, admit that they know very little; yet there is a great deal more to learn. Also, the ethologists are devising many new technical terms that are sometimes hard to understand. Aside from my university training as a biologist, my own interest in and knowledge of animal languages has come from two years of studying wildlife in the jungles of Panama and about thirty-six years of living on ranches and farms, where, with great interest and enjoyment, I have watched, studied, and learned to understand the tame animals and birds we have always had around us. I have gone on countless hikes into the surrounding wildernesses to see and learn about the creatures of the wild.

I must frankly admit, however, that the word *secret* in the title of this book is a very good word for much of the still unknown languages of animals, and that we are only peeking in here and there to see and try to understand the most obvious signs they give to each other, and to us. Since even the scientists admit much lack of knowledge, it would be wise for us to always keep very open minds and to be ready at all times, when we talk about what animals are saying and what we believe are the natures of their communications, to admit we can be wrong!

Nevertheless, I feel that this book will lead you on to many fascinating trails, and you can use it to help to begin to understand about the vast and interesting subject of animal languages.

The chart entitled "Animal Communications," shown on page 8, is based on similar charts by ethnologists, but it is my own attempt to make the knowledge of animal languages easier for amateurs to

ANIMAL COMMUNICATIONS

Miscellaneous Communications

- Escaping other dangers
- Long-distance signals
- Animal-Human communication
- Animal-Plant communication
- Training young
- Migration
- Mimicry
- Camouflage
- Play
- Indecisive action, or freezing

Escaping Enemies

- Escaping a larger member of same species
- Signaling retreat or acceptance of dominance — same species
- Signaling danger from carnivores
- Escaping carnivores

Attacking

- Attack by mobbing
- Signaling mob action
- Attacking similar but different species
- Attacking within same species

Social Communication

Slave making	Nesting	Building city or community
Migration	Foraging	
Attacks by packs or other groups	Trail making	Social defense
Teaching young	Feeding	

Fixed Action Displays

- Male attracting a mate
- Female accepting a mate
- Guarding territory
- Guarding young

A Scientific Approach

understand. In the paragraphs that follow, I will give an example of each part of the chart, so that you will understand it better and also be able to use it as a guide to the information given later in this book. My breakdown will start with Fixed Action Displays and proceed counterclockwise around the chart, ending at the center with Social Communications.

FIXED ACTION DISPLAYS

Fixed action displays are habits that over hundreds of centuries have become deeply ingrained in each species of animal. These behaviors are brought into play under particular circumstances, and usually relate to starting or maintaining families. Other fixed action displays occasionally will be observed when animals are resisting attacks or mobbing a carnivore.

Some species are much more flexible about these displays than are others, and some displays are much less visible than others, although just as important to the species. We have to be very careful in identifying a fixed action display, because there are behaviors that look like these displays but that in reality have other meanings. For example, the American anhinga, a large shore and water bird found in the southeastern United States, has a very snaky neck and beautiful, silver wing patches. It swims underwater but must be careful not to get too wet. In order to prevent this, it often comes ashore, lifts its wings high, and shakes them vigorously. Most people, seeing this for the first time, think it is some kind of display being put on for the benefit of other members of its species. But this is not so at all. Watch carefully when you see something that looks like a fixed action display to see if another bird or animal is watching and responding to it.

Fixed Action Display of a Male Trying to Attract a Mate

The redwing blackbird male performs a high-intensity fixed action display over a marsh when attempting to attract a female. It claims a special, small marsh area as its territory, and puts on its display both to attract a female and to drive away other males. Each male redwing blackbird has a red-and-yellow epaulet of bright-colored feathers on the forewings. These epaulets are hidden most of the time, but when it displays, the covering dark feathers are lifted off the epaulet, and it shows in all its glory as the bird flies above its territory, singing loudly to the female, its body bent in a peculiar way. This fixed action display is done over and over, until a female is attracted to join the male and they start building a nest in the thick reeds of the marsh.

Fixed Action Display of the Female Accepting a Mate

Usually a female puts on a much less visible display when she accepts a mate. It can be as quiet as simply joining him, but a typical female response when she accepts a mate is to start putting together a nest or to squat low on a branch, showing she is ready to mate.

Guarding a Territory

Birds usually sing when guarding a territory, but they also prepare to attack. Mammals usually leave scent trails or special scent markings on bushes or trees, as when a bear scratches a tree high up on the trunk and marks it with his hair and scent. It is dangerous then for another bear, especially a male, to enter this scent- and scratch-marked territory. Fence lizards guard their territories by displaying their blue throat patches and bobbing up and down. Always, this simply says, "Stay away from this area that belongs to me!" or "Come to me!" to a possible female mate.

Guarding Young

There are several ways of guarding young. One is simply to attack or show signs of attacking another creature that gets too near to the young. King birds are especially vicious and use loud chattering calls when they attack, though even a king bird will be careful not to get too close to a hawk and to always attack from above. The loud call of warning, and then the attack, is a fixed action display common to attacking patterns at the edges of a territory.

In another fixed action pattern, in order to guard the young, an animal lures an intruder away by acting injured, like the killdeer I mentioned earlier. Another time, a long-billed curlew lured me away from her nest by dragging her wing on the ground as if it were broken, so that I followed her to see what was wrong. When she finally got me far enough away, she gave a cry of triumph and flew high

A Scientific Approach

above me, but away from her nest, until she knew she was out of my sight. This fixed action pattern, developed over many centuries, says, "Look! I am hurt! Come follow me."

ATTACKING

Animals attack either to get food or to drive a competitor away from food. In the first case, the attacking animal is usually bigger, stronger, or better-armed than the one attacked. It simply says, "I am going to get you!"

Attacking within the Same Species

This attack is rarely meant to kill; rather, it is intended to show that one animal is bigger and stronger than another and wishes to drive the other away. The "peck order" in the chicken yard is an example of this (see page 47). The attacker usually ruffles up its feathers or hair to show that it is about to attack.

Attacking Similar but Different Species

This attack most often simply drives a weaker species away, as when a jay drives a sparrow away from some food. The jay's feathers ruff up and, most important, its topknot moves high to show it is about to attack. It is saying, "That food is going to be mine!"

Signaling Mob Action

If a carnivore (like an owl or a wildcat) is spotted where birds can get at it to mob and harry it, usually one of the bigger birds (such as a jay) sets up a kind of war cry, yelling as loudly as it can, "Jay-jay-jay!" This attracts other birds of all sizes to join together for an attack. It is saying, "Enemy! Let's bother it!"

Actual Attack by Mobbing

In a mobbing attack, birds dart down at one or more carnivores, trying to bother the enemy as much as possible. Different species often work together, but sometimes a mob is made up of only one kind of bird (as on page 156). If a small owl is found in daylight on a tree branch, birds may actually rush close enough to kill it with vicious pecks. But a larger carnivore, like a great horned owl or a wildcat, must be handled more carefully, as it can counterattack, suddenly plunging out at the attacking mob and possibly catching a bothersome jay. Then the fun is over!

MISCELLANEOUS COMMUNICATION

Long-distance Signals

Some animals can communicate over very great distances. The messages may be mating calls, or they may have other meanings, such as the calling together of a pack, herd, or pod. For example, wolves have a powerful howl that can be heard for many miles in the wilderness on still fall or winter nights. The pack leader begins to call the wolves together for a hunt, and the others answer him from as far as ten or more miles away. Giant female saturniid moths, such as the luna or the polypheus, can send out on the wind pheromones (chemicals that attract and stimulate other animals) that may reach males several miles away and call them for mating. You can test this yourself if you can find the large cocoons of these moths. Isolate the males as they emerge (they have large, feathery antennae). Mark them and take them two or three miles away and free them, keeping the females in cages at the starting point. See how long it takes the males to reach them.

Greatest and most wondrous of all long-distance signaling is the singing or crying of such huge creatures as the humpback and fin whales, which can send their calls through the deep ocean for literally hundreds and even thousands of miles, probably to call each other together for meetings. A pod of humpback whales coming together to sing is a tremendous event to listen to. These songs have been put onto tapes by scientists, but no one has yet decoded what the whales are saying.

Animal-Human Communication

Anyone who has owned a pet cat, dog, parrot, canary, or even snake or lizard, knows that after a while the animal begins to know the meanings of certain words, and that humans begin to learn what animals mean by the sounds they make and the ways they move. But just as most human words are too hard for animals to understand, so some of the sounds, smells, and actions of animals are still too difficult for us to decode. Some creatures, like bats, employ sounds that are beyond our range of hearing, and some, like dogs and deer, are able to catch scent meanings on the wind.

Animal-Plant Communication

In order for some plants to reproduce, their pollen must travel from one flower to another. Some of this cross-fertilization is accomplished by the wind, but birds and insects play an important role by carrying the pollen on their feet and feathers. By the colors of their petals, and possibly the smell of their nectar, flowers such as the monkey flower and the honeysuckle "call" to hummingbirds, whose long, pointed bills are especially suited to taking the nectar from the flowers' deep, narrow throats. These flowers also attract the sphinx moth, whose long tongue can reach the nectar. Both plant and animal are well served in this relationship, because, while the bird or moth feeds at the flower, pollen sticks to its body and is then carried away and deposited on other nearby flowers. Bees and flies are also called to particular flowers and aid in their cross-fertilization.

Training Young in Survival Skills

All the higher animals try to teach their young how to live, first by playing with them to strengthen their muscles, then, if they are carnivores, by showing them how to catch small game, and bigger game when they are ready. An important part of this education is showing the young ones how to catch other animals without being bitten. If you have a chance to observe such teaching, you can almost hear the adults saying, "Look out, it will bite!" or "Run fast and catch that one!" The young also have to be taught how to avoid dangerous animals, such as man.

The smaller, vegetarian mammals have to be taught to behave defensively, including, if worse comes to worse, how to "play dead." By nosing her young, the mother rabbit expresses all kinds of commands that you can put into words if you catch the feel of what she is saying.

Some animals teach their young using very primitive methods. The male stickleback fish guards both eggs and young in a kind of brush cage he makes to protect them from predators. His main teaching activity is showing them how to eat plant and insect food. They soon learn to follow his actions. When big enough, they are on their own.

More highly developed methods of teaching are found in animals and birds that like to play with their children. A mother and father red fox, for example, are very solicitous of their young ones' needs during the first days after birth, but when the little foxes show signs of playing, the teaching begins. First, dead mice and rats are brought to the young and torn up so they can be easily eaten. Then, as the kits get older and more sophisticated, the parents bring young live mice, which run and stimulate the kits to chase and seize them. When the young foxes lose their prey, the parents give them lessons in how to capture mice correctly. The kits usually catch on quickly and are ready for the next lesson: bigger and more dangerous creatures, such as rats, that can give bad bites if handled wrongly. The mother fox shows the young how to seize the rat quickly by the neck and get a good grip so there is no biting back. If the kit does get bitten, it is very anxious to learn better next time!

Next, the kits are taken out on trips through the woods and shown how to find mouse and rat holes, and later, rabbit and woodchuck holes. They are taught to wait quietly in front of a hole, lying perfectly still but with hind legs braced for a quick lunge when the mouse or rat sticks its head out. They are also taught to follow trails by smell and to detect the smell-boundary scratches and scent posts on trees that mean other foxes don't want them in that territory.

One of the most difficult lessons is how to avoid man, the most dangerous animal of all. Young animals quickly learn how to elude human intruders, who often carry guns and lay traps. The animals use their sense of smell, that special language that has been lost to man. By observing animals very closely after they smell something interesting, and by following them on foot or through binoculars, we may be able to learn more about these secrets that are carried on the wind or that they place on rocks and trees.

There will be more about animal-human communications in the chapter on domestic animals.

Migration

When animals migrate, they use their communication skills not only with each other, but to sense patterns about the weather. When the weather begins to get cold, many birds and some powerful flying insects begin to migrate south, and mammals may make southward movements or at least come down from the high, cold mountains into warmer lowland areas. Sometimes great groups of similar bird species move together in migrations, but these groupings are apparently casual and temporary—as if just for the pleasure of companionship—as opposed to real social migrations, in which a band or flock acts together as one unit following a leader or leaders. Thus warblers may flock together with sparrows and vireos on trips south, all cheeping to keep in contact, but without leaders directing them. It is difficult to get their high-flying calls on tape, but we can begin to catch some of the special ones they give when they first come together.

Almost all flying is done at night, as songbirds are afraid of carnivores during the day, and owls rarely fly very high at night to search for migrators. When storms begin to come, the cries of the small birds become cries of fright, and they try to get down into the trees as soon as they can. If you lie on a high porch at night during the migration season with a telescope or binoculars, you can watch the flight of birds across the surface of the moon and perhaps identify the larger ones. Watch for flocks that fly in ordered groups (geese) or those flying in scattered groups (small birds).

Mimicry

Animals mimic other animals, and sometimes plants, in order to hide themselves from their enemies or to make a predator think they are much more dangerous than they really are. Thus, certain hawk moths and catocala moths have big eyelike spots on their underwings. When at rest on a tree, they fold their upper wings over lower wings, by mimicking the bark they appear to be part of the tree and so fool predators. But suppose a bird becomes suspicious or has just seen the moth light on the bark and pecks it on the back? Instantly, the hawk moth spreads its front wings widely, exposing those big eyelike markings on the underwings. The bird sees the markings as the eyes of a dangerous large animal and startles, allowing the moth to escape. In the first case, the hawk moth says, "This is just bark, so there is nothing to eat here!" and in the second, "Look out! I am something dangerous!" Both behaviors help it escape its enemy. In both cases, it is practicing mimicry.

In another common case of mimicry, an insect mimics a much more dangerous insect. A fly that is yellow and black carries the colors of a yellow jacket or hornet. It tells potential predators, "Look out! I'm dangerous!" But it really isn't. Stick insects have very fine mimicry, especially when they stand still. They are saying, "I'm only a stick. Don't bother with me!"

Camouflage

This is similar to mimicry, but has more to do with merging colors with the environment so a predator cannot see the one who is hiding. Many creatures, such as mice and rats and rabbits, crouch low among grasses and bushes to try to hide, and some predator insects, like mantids, pretend to be leaves in order to catch unsuspecting prey.

Play

The language of play is found among the more advanced species of animals, not only among the young but often with the adults, too. It is mostly a very simple language. A kitten, for example, starts playing with a ball, and very soon other kittens catch the call to join in. But games can become more complicated, and the more intelligent animals, such as ravens, crows, and foxes, may play more elaborate ones. Ravens have been known to play a game with a stick, tossing it to one another in spiraling movements while they fly. Imagine one shouting to another, "Here, catch this if you can!" More examples of play will be given later.

Indecisive Action, or Freezing

Indecisive action can occur when a bird or other animal does not quite know what to think about another. The eastern phoebe, for example, is a small brownish bird that lives at the edge of woods, around bridges, and in gardens. In the spring, the male announces its presence by a "Twh-t" call, low to the ground, or a "Phee-bee" song in the trees. The first call is usually to establish territory, while the "Phee-bee" song is to attract a prospective mate for nesting and egg laying. Bird students have noticed that the "Twh-t" call, given mostly when the bird is foraging for food, may become very indecisive, and that suddenly the bird will fly high in a tree to sing. The indecision is

A Scientific Approach

the result of two separate urges, the urge of hunger and the urge to find a mate.

Another example of indecision is when the phoebe sees an enemy, such as a hawk or a cat. The bird hesitates because it does not know whether to flee or freeze. If it stays perfectly still, it may look like part of a tree and be unnoticed by the carnivore. If it flees, it may escape, but it may be captured. If it freezes, the phoebe is saying, "Now you can't see me, so go away!" If it flees, it is saying, "I am going to dodge among the leaves of the trees, and you can't catch me!" If the predator attacks, it is saying, "I am going to catch you!"

ESCAPE FROM ENEMIES

We have already seen examples of escape from enemies in the sections on mimicry and camouflage, but here we will consider the phenomenon more fully. Escape begins with the knowledge that danger is approaching and may have to be avoided and can include warning the family or flock, or even other species of friendly animals. Often creatures hide before the dangerous animal or man can get close enough to see them. They dive in among the plants and look like them when a hawk soars overhead or a fox comes searching. In a way they are speaking to the plants and saying, "Hide me!" In another way they are speaking to the hawk or fox, saying, "You can't see me now!"

Escaping a Larger Member of the Same Species

Running or flying away from a bully is one way to say, "I don't want to fight!" In the chicken yard, all the chickens say this except the biggest and strongest rooster. Different animals have different ways of escaping and different sounds of warning.

Retreat or Acceptance of Dominance

A young animal or a smaller animal may retreat from a larger animal of the same species, or accept dominance by such an animal. In a pack of wolves, for example, a young wolf will lower its head and tail and crouch before the dominant leader of the pack to signal submission and accept being bitten without returning a bite. It is saying very clearly, "You are the boss!" This behavior is common to most doglike animals. A young bull in a herd of cattle or buffalo mainly stays out of the way of the dominant bull.

Signaling Danger from Carnivores

The white, raised-tail flag of a deer or antelope is usually a signal of danger. The nose points in the direction from which the danger is coming. Sparrows and other birds use white streaks on their heads to point in the direction of danger and may also give a cry that means danger from that direction.

Escaping Carnivores

There are four main ways of escaping from carnivores: outrunning or outflying them, outdodging them, quickly hiding from them, and diving into a hole too small for them. Any one of these, if successful, is a triumphant cry of "I got away!"

Escaping Other Dangers

A sudden heavy rain or snow can quickly create a disaster for many animals, especially small ones. Tiny squeaking or rasping cries have been heard from insects at such times, suggesting that they may be shouting warnings such as "Get under cover quick!" to their mates. Spider mothers may try to protect their eggs with their bodies or to drag the eggs under a leaf or rock. Angleworms come to the surface of the ground to escape drowning.

SOCIAL COMMUNICATION

Complex forms of communication are found among socialized, communal insects, such as ants, bees, wasps, and termites, and among social animals, such as antelopes, wild horses, wolves, and beavers.

Attacks by Packs or Other Groups

It is mainly hunting animals, like wolves, or driver or slave-making ants, that attack other creatures. They are essentially trying to say, "Here we come, and you cannot escape!" If a pack of wolves can convince a deer or moose that all hope is gone, it is much easier for them to kill it. Driver ants form armies that march over the ground in such numbers that insects and spiders as well as mice and rats feel helplessly surrounded and die with hardly a fight. Warfare can occur between large ant colonies, not only when slave-making ants attack, but also when two nests of ants are so close together that they begin fighting for territory. When such fights start, tiny cries or legs scraping together or against the body may call for help or goad the warriors to victory. Ants may leave pheromones on the ground to call other ants to come help in a battle or to direct ant armies to places to fight. This action may be translated as the cry "Come quick! We need help in a hot fight!"

Trail Making

Colonies and cities of ants are great trail makers. A single ant may have several kinds of smell markers or pheromones in its body. It can leave molecules of scent along a trail that it is following or help make trails through grass or forest areas or on the floor of a desert. These pheromones give various signals to other ants, such as, "Good food down this way!" or "Enemies ahead!" You can often tell what messages were conveyed by the ants' subsequent actions.

Beavers make smell trails to show where good tree bark is to be found. Deer may mark out a smell trail in winter to show where a herd is trodding down snow to keep an area open for feeding and escape from predators. Wolves leave trails leading to the best hunting areas.

Social Defense

A beaver house is made by one or more beaver families as a fort against wolves and bears. Ants form lines of defense around their cities to prepare for war. Social wasps defend themselves from many enemies by putting nests in hiding places and even decoys to lead enemies in the wrong directions. Prairie dogs often put sentinels on the edges of their towns to warn of the approach of a predator.

Teaching the Young Social Behavior

In an animal society, mothers and fathers constantly try to teach their young how to behave. They show them the territorial boundaries beyond which they should not go because of danger from predators. They may even teach them how to act toward their elders. Most teaching is done by example; parents nudge or even slap or nip their young when they do something wrong. When a flock of geese fly north in springtime, the young are usually placed in the middle of the flock and watched over by the elders to see what they do and how they follow the leaders. Thus they learn the way of the flock and what to do about enemies. Insects in colonies learn by experience to take on more responsible jobs; usually the youngest members do the simplest work, such as cleaning out the hive or nest; they graduate to defense or easy food gathering, finally becoming scouts or trail leaders. Touch and smell, rather than sound or what the eye sees, may be most important in such learning.

Feeding

In ant, bee, or wasp colonies, the young grubs or larvae are usually kept in the comb cells or in special areas where they are fed, often mouth to mouth, several times a day by the mature bees and ants. Somehow, in ways we need to find out more about, the young larvae let the grown ants and bees know when they are hungry, possibly by simply opening their mouths.

Among bird flocks or mammal packs it is usually easy to know when it is feeding time, as the young mammals yell for food, though when danger is near, they have to be shushed up very quickly. Young birds in the large flocks yell when they get hungry and show large, gaping, red mouths that indicate "Feed me!"

Slave Making

This social feature is found entirely among ants. Only a few species form slave-making armies that attack other ant cities or nests, kill the adults, and take away the young larvae to be raised as slaves for the warlike city. Larvae taken in this way pupate and emerge as workers perfectly adapted to care for their masters, with no memory of their old home. They work willingly at all the menial tasks of the city and even fight and die for their new home when it is attacked. The fighting teeth of some kinds of slave-making ants are so particularly adapted for fighting that they cannot gather food for themselves and must depend entirely on their slaves. When the slave-makers attack, though, they are very efficient!

Nesting and City Building

Elaborate structures are built by a number of insects and animals, including ants, wasps, termites, beavers, and birds. It is obvious that when such structures are built some kind of communication must happen. (Chapter 8 of this book discusses the social communication of bees.) This is one of the most fascinating challenges for the student of animal languages, as a great deal more is yet to be discovered. We can mainly ask questions in this area rather than know what actually happens.

For example, who organizes the building of a termite city in wood, an ant city where large rooms house fungus gardens that supply the city with food, or a hornet nest hanging in a tree? These questions have not really yet been answered. Is it the queen who does this in each case, or is there some kind of collective intelligence doing the designing and ordering? We do know that ant and bee queens may actually have smaller brains than workers.

We think that the oldest beaver, probably a male, designs and organizes a beaver colony along a creek, although further research must be done. Beavers signal danger by slapping the water loudly with their tails, and they are often heard mumbling and grumbling or even giving fighting sounds inside their beaver huts, but we are not sure what they are saying. One thing we do know is that an older and bigger beaver usually shows the others how to build a dam. He does it by actions rather than sounds, and he may slap or bite a younger beaver who does a poor job, and make him do it over.

Colonies of nesting birds may protect each other by yelling and screaming and even attacking an animal that enters their colony, but they do not work together to build the colony as beavers do.

Colony or City Structures

The hives of wild honeybees are usually made in hollow logs or narrow caves, where the beeswax is fastened to the wood or stone. The bees secrete the wax from their abdomens and chew it until it is firm enough to be used in constructing the hollow chambers in which honey can be stored and bee grubs reared. The building of such a colony is started in early summer by a single queen, whose fertilized eggs hatch workers who help her build the hive. Instinctive knowledge is born into the first workers, but the queen must do some leading and directing.

You can make an artificial beehive with glass sides using a wooden frame to support the glass. The frame should be about two inches thick, and should form a rectangle measuring 10×16 inches. In one corner, leave an opening of about $\frac{1}{2} \times 3$ inches to allow the bees to come in and out. The glass is glued to this frame with honeycomb, which, along with a queen and several dozen workers, can be obtained from a commercial bee company. The glass should be covered with a dark cloth except during times of observation. Of course the hive will soon be outgrown, but even a few days of watching can give you much knowledge about the lives of bees. You will see that most communication among bees is through touch, smell, and sight. Close watching will gradually give you a feeling for the bees.

The artificial bees' nest will also do very well for wasps, but they are more dangerous. A new young queen must be captured in the springtime and put into the combined cage and nest. Wasps can be fed pieces of meat and nectar, and you can watch the colony grow between the glass walls. A larger cage can be attached to the artificial nest with a passageway between the two enclosures. Potted plants should be put into the larger cage to give a natural-looking enclosure, and live insects provided to the wasps as food.

Termite and ant communities can also be watched through glass cages, with narrow wood pieces or earth separating the two pieces of glass. Again, doors can lead to larger enclosures, where the community activity as a whole can be observed and the insects kept well fed, with meat and grain for the ants, and wood for the termites. Water and food must be continually furnished.

For communication, touch, smell, and hearing are more important to ants than sight, though some have large eyes and can see well. Pheromones, or smell molecules, are used continually by many ants. Experiment by putting two kinds of ants together and with putting ants into termite colonies to see how they react. You will often find that one kind of pheromone used by ants means danger, another is a sign of nearby food, another a command to attack an enemy. Fascinating things can be learned by careful watching of these artificial towns.

Summer vacations are good times to visit beaver colonies in the western mountains. Sounds and smells are most vital for communication among beavers. They slap their tails on the water to warn of danger, and use castoreum, a smelly substance, to pass about knowledge as to where new food can be found or if danger is coming.

Migration of Social Animals

Migrating flocks of birds fly to the North every spring and fly back to the South every fall. It is interesting to watch their social life both in the ground and in the air. Most small birds fly in loose flocks of several kinds, simply going from north to south or south to north according to season. But some small birds, like the goldfinches, fly in more organized flocks and you can watch to see if there are any leaders giving signals to control a flock and its direction of flight.

Larger birds like the geese fly in flocks of about one to three families, usually with a wise old female goose in the lead to give the signals. I have watched unified flocks of geese in the sky, but these are generally one- or two-family units rather than great unified flocks of hundreds. Small flocks of mergansers often fly in groups up and down the big creek that lies near our farm house. They probably also are family members. Very large birds such as the cranes may travel together in large flocks that sound as if they are constantly communicating back and forth. I have seen huge flocks of sandhill cranes come in to rest and play in the marshlands of southeast New Mexico, preparing for their northward flight of migration to the plains and marshlands of western Canada, but it was not possible for me to tell whether they were groups of families like the Canada geese or were one huge synchronized flock. Some males were doing the high-jumping dances by which they attract the females, and I often heard their shrill but almost roaring "Garooo-ah-ah-ah-ah," like a Scottish gathering of the clans.

Some migrations are up and down high mountains rather than from south to north and back. Herds of wild burros of the southwest go up into the mountains in the heat of summer and down into the deserts in winter, the wild and raucous braying of the leaders directing the herds to new areas to gather food.

Probably the greatest migration of insects in North America is that of the monarch butterflies, large brown and yellow-red insects that fly each fall in great masses from north to south; even greater numbers spread out from south to north in the spring. These large butterflies are poisonous to most birds and so are generally left alone when flying. In Pacific Grove, California, next to the sea, they hang in myriads in the conifer trees. They must communicate mainly by the sight of each other in flight, but what keeps them together in such masses in the trees is not yet fully known.

Foraging

Cooperative foraging is very important for social animals of all kinds. A wolf pack leader may send out scouts to find deer or moose for the pack to hunt down. How he does this we do not yet quite know, but by bark or growl different meanings can be given. Foraging for antelope is mainly controlled by sight; they look for plant food on the Great Plains and other wide-open places like sagebrush flats. Scouts who find new grass come back in sight of the herd and indicate by their physical attitudes, primarily by pointing with their noses, that good grass or herbs are near.

Ants usually signal by sight or smell where available food is located. Army ants, who are sometimes found along the southern border of the United States, are completely blind; they send out scouts who drop pheromones to show where there is good hunting. Harvester ants, on the other hand, use sight almost exclusively; when groups go foraging, they watch for the leading foragers, who indicate by their excitement that new grain has been found. Fire ants use both sight and smell, but mainly the latter, and their sense of smell is so powerful that they can detect food from great distances. All such sights and smells are saying, "Hurry, here is food!"

3
Domestic Animals

This chapter concentrates on the languages of a few of the familiar domestic animals. The languages of such animals are strongly influenced by man, and in turn the animals often use the learned parts of their languages to influence man.

DOGS

When I open the door of my house in the morning and go outside to start the chores, our two dogs, Billy and Tad, rush up barking to greet me. It is a glad, happy bark, not the sharper bark of alarm and warning given when a stranger comes. If I pet one, the other jealously shoves in so his head will come under my hand, too. Then, as I walk toward the barn, both run in front of me, turning their heads to look back and making a distinctive deep growling whine that expresses several emotions, including joy, pride, and an invitation to play. It would be impossible for me to put all of it into words.

There are so many differences among dogs as to character, and their characters are influenced so much by the various kinds of masters they have, that not even a large book devoted to the subject could cover all these variations. Here we can give only some of the basic ideas of dog language.

In my own experiences with dogs I have progressed from lack of fear through fear and back to confidence again. As a nine-year-old boy I was bitten severely by a police dog, and for years after that I was afraid of dogs. They all sensed this and gave me a bad time, approaching with hair raised and growls in their throats. Later, by an effort of will, I managed to conquer this fear and learned to express from my heart a feeling of love and interest in all dogs. The change in dogs who approach me has been truly remarkable. Now it takes only a minute or two of quiet talk and slow movements for me to make good friends with most dogs.

Dogs are extraordinarily sensitive to the feelings of the human beings around them, and they may even detect evil in a person whom the master considers perfectly all right. A dog will often drop its tail and look miserable when told to be friendly with someone he instinctively dislikes. Dogs also are likely to absorb the characters of their masters. The dog of a man who is a bully may also be a bully. Dogs who are kept chained and severely disciplined tend to be very vicious toward strangers.

To understand dogs we need to know something of their ancestry. The chief ancestors of most dogs were probably closely related to the present-day jackals of central Asia. They were domesticated around twenty thousand years ago. Later, as men began to advance into the far northern lands, the domain of the wolf, the jackal strain was here and there crossed with that of the wolf.

The attitudes toward man of the jackal strain and the wolf strain are quite different. The jackal is a slavish follower; the wolf can become an independent companion and friend. The jackal dog is inclined to be friendly to all men. The wolf dog is much more aloof and usually chooses one man as its pack leader. In jackal dogs the loyalty to the pack is not very strong. In wolf dogs it is very strong, and the wolf dog transfers to its human master the loyalty that at one time would have gone to the pack leader. The jackal-dog attitude toward the master is more that of a puppy toward its mother: it remains childishly dependent on the human master, even after it is adult.

The great majority of domesticated dogs have mainly jackal blood. A few—such as the Alsatian, the chow, the husky, and the German shepherd—have chiefly wolf blood. As you study dogs and the way they communicate, keep your eyes open and notice the difference between the two main types. You will, for one thing, notice that some are friendly toward all humans and recognize no particular master. These are jackal dogs. At the other extreme is the very independent pet who also obeys no one master, yet is usually friendly toward one

The stiff tails and erect hair indicate that these dogs are ready to attack each other, but if one concedes by lowering his tail, the fight will not start.

person though hostile toward nearly all others. The dog who makes the best pet is somewhere between these two extremes, loyal to one person above all others and protecting his master's family, but friendly also toward other people if properly approached.

A dog has much better smelling and hearing faculties than we have, but its sight is usually not as keen as ours. This is why sometimes your own dog may come rushing toward you, barking and growling, if the wind is blowing from it to you. It sees you as a kind of blur and does not recognize you as master until close enough to smell you or to hear your voice.

Watch two strange dogs approaching each other, and notice the various ways they communicate. If one is a puppy, it turns over on its back and makes water, frantically waving its tail. The older and larger dog approaches and sniffs over the puppy, also wagging its tail. A male dog would no more think of attacking a puppy than of attacking a female.

If the two dogs are both males of nearly equal size, and strangers, they approach more slowly, with legs stiff, hair on backs raised, rumbling growls in their throats. They come to a halt close together, with nose to tail, and sniff at each other's tails. This is as strict a rule as anything at a diplomatic reception. Meanwhile both dogs keep their tails absolutely erect as a sign of their courage. If either tail should begin to drop, it would be a sure sign that this dog was becoming afraid. He would probably soon turn and run at full speed, with the other dog snapping at his heels. If, however, the tails stay up, there is usually a prolonged period of growling and slow circling, each dog trying to outbluff the other. Then sometimes one begins slowly to wag his tail. This is a sign he wants to call a truce and play, instead of fight, but that he will fight if necessary. When the second dog wags his tail, too, the two run off for a romp.

If a real fight is going to start, the first sign is the drawing back of the lips into a horrible grimace. The second sign is usually that each dog suddenly scratches his feet vigorously on the ground. After this they close in on one another with loud growls and yells, and the fight begins. In the fight, each dog tries his best to meet with his own teeth the teeth of the other dog. The teeth act like shields or like fencing swords. Really skillful fighters will counter teeth with teeth for some time before going in closer. Except with certain breeds—such as the bull terrier, who is trained to kill—these fights rarely end in death. When one dog feels he is being beaten, he may suddenly expose his neck to his enemy. You would think this would mean the end, that he would be seized by the throat and choked to death, but this rarely happens. The other dog growls horribly and stands within an inch of the unprotected throat, making no attempt to grab. This is because the first dog has begged for mercy. The chivalrous rule among dogs, as among wolves or jackals, is that a plea for mercy must be observed.

However, the dog who has turned his throat dares not move for some time, and he stands still with head lifted. Finally the victor decides to mark this place of his victory and makes water on a nearby post or tree. The instant his back is turned, the other dog runs; and, from that point on, the losing dog always lowers his tail when he meets the winner and either runs or tries to be friendly.

The situation is sometimes changed if it is a much younger dog who is licked in a fight. The young dog gets older and stronger while the older dog gets older and weaker. Finally the younger tries a fight once again and wins.

The attitude of the male dog toward the female is very different. He is usually as friendly as can be, wagging his tail and trying to get her to play. She, however, may resist his advances by sudden growls and snappings at his face. At such signs of disfavor, he draws back with a most pained look, as if to say, "Oh, come now, let's be friends!" Gradually she may thaw toward him and become more friendly, leaping up and pawing at him with her front feet. Sometimes, though, she remains grumpy so long that she discourages him completely.

Both male and female dogs leave scent signals wherever they water. When the female feels the mating urge, she goes on a trip around the countryside, leaving her smell signs where male dogs will find them. Then she confidently awaits their coming to find her. The male dog leaves his scent signal not so much for the female as to warn other males by the strong odor that he is a big, fierce dog and that this is his territory.

Most dogs do not mate for life as the wolves do. The male mates with many different females, as the jackal does. So there is little family life, and it is less likely that dog packs will have the close cooperation found among wolves. The huskies of the far north are an exception because of their strong wolf blood.

The sounds that dogs make convey different meanings. The whine of the puppy is a demand for food. The yelp is a sign of fear or of pain. The bark of a dog has a feeling of interest and excitement and warning. The annoying, shrill bark of a small dog has in it something of fear. The deep, almost savage bark of the big, powerful dog sends forth a threat. When dogs bark as they play, the sound is full of joy, but the sharp, insistent bark of dog to master usually means it is begging for something or wants some action: "Get the ball and let's play!"

The dog is even more expressive in sending signals with his body. It wiggles all over with joy when it knows the master is going to take it for a romp. It may seize its master's pant leg to hurry him along. It may whirl around and around with excitement, barking furiously. Some dogs will bring a ball or a stick when they want it thrown. A common signal from a dog who wants to get in or out of a house is going to the door and scratching at it. If this is forbidden, it will learn to stand there and whine, or just stand there with big eyes looking sorrowfully at the door.

Our two dogs spend hours playing with each other. If Tad wants to start a game, he nips Billy in the rear and runs away, with Billy in close pursuit. But if Billy isn't interested, he lowers his head between his paws and growls. This means "Go away! I want to rest." Sometimes the play gets a little too rough and one of the dogs growls a protest. The other crouches, with his front body low and his tail end up in the air, barking loudly, as if to say, "Aw, come on! Don't be mad just because you got nipped a little too hard!"

A younger dog often watches an older with great care. If the older dog barks, the younger barks, too. If the older dog growls, so does the younger. By this copying, the younger dog tries to act like an adult. Young dogs who have been brought up completely isolated from older dogs show both more nervousness and more stupidity when faced with emergencies because they have not learned how to meet them.

A good trainer teaches a dog many new words, not only those for it to expect from the master, but those for the dog to use in reply. A hound is taught to come for help and to retrieve. It learns to stay with one kind of game instead of being distracted by others. Some of the signals sent back by a hunting dog to its master are instinctive, however, as when it barks with great excitement and eagerness beneath a tree in which a wild animal is treed. And many bird dogs instinctively point with their noses, or with one leg uplifted, when they sight a game bird.

The Eskimo dogs, or huskies, are much like wolves in their reactions and their methods of communication. They form packs of five to ten individuals, with each pack usually attached to one master and one sled. Each pack has its own special territory, which it defends ferociously, and its own leader that the other dogs in the pack obey because of its superior strength and wisdom. Young huskies, who have not yet learned the territories, are always being chased off or nipped by their elders.

The numerous ways in which dogs talk to men or to other dogs—the bark of anticipation, the whine of pain seeking help, the howl of loneliness when the beloved master is gone, the hankering whine of a male trying to woo a female—can only be touched on in this book, and many must be left out because of lack of space. Study your own dog or the dogs of your neighborhood and observe how their feelings are expressed in their voices and in their movements. You will soon be able to understand the language of dogs, particularly if you feel kinship with their emotions. With love comes understanding.

Domestic Animals

CATS

Just as with dogs and men, some cats are stupid and some intelligent. Geniuses are no doubt capable of some really remarkable things. However, here we are only going to talk about regular run-of-the-mill cats and how they express themselves.

Cats have no pack language like dogs, and little interest in obeying the commands of men—though there are exceptions. A cat is primarily interested in good food, a comfortable place to sleep, and other cats—particularly those of the opposite sex. It uses different tones of a meow to obtain what it needs. It uses a deep, humming purr to tell others when it is completely happy, a guttural growl to express displeasure or anger, and, if a male, the most terrible variety of caterwauling, moaning, growling, and screaming when it wants to tell another male cat it doesn't like him.

A cat leaping at a mouse signals other predators to stay away.

The Secret Languages of Animals

A cat is a solitary hunter. When it hunts a mouse, the wriggling tail tip, the slightly swaying shoulders, the tense crouch, and the glaring eyes tell all other cats and carnivores to leave that particular mouse alone. If robbed of the mouse, it lets out an outraged squall. The cat's habit of playing with a mouse it has captured but not killed is cruel to the mouse, but it is very useful for a mother cat in teaching her kittens to hunt. The clumsy way a kitten misses its first pounce shows how much it has to learn.

The cat has never become as completely domesticated as the dog. Many dogs will let a child pummel them without protest, but a cat soon reaches the end of its patience and turns with a little meow or growl of anger and uses its sharp claws and sharp teeth to get away. This meow, given with a rising inflection of annoyance, is a milder warning than the deep growl and hissing of a thoroughly enraged cat.

The cat's final stand against injustice and real danger is made with every hair standing on end and the tail sticking straight up in the air. The ears are laid back against the head and the lips drawn high to expose the white canines. This menacing figure, combined with a ferocious hissing and snarling or moaning, is often sufficient to keep a small dog or even a medium-size one from attacking, unless it is thoroughly reckless or has had no previous experience with those flying claws and sharp teeth.

Two male cats preparing for a fight do not act this way, however.

Tomcats display aggression in an attempt to intimidate one another.

The primary objective is for one to bluff the other into running. The hair is lifted, though not standing on end, and the tails lash back and forth in spasmodic jerks. If the tail lashing increases in speed, we know one cat is about to attack. Ears are laid back and faces are set in snarls of rage, but the main bluff is carried on by sound, particularly a deep moaning, interspersed with spitting, fearsome wails, and savage growls.

If neither cat runs, one will finally grow impatient, lash his tail furiously, and suddenly rush his enemy, his belly low to the ground. The reason for the low rush is that, as soon as the battle is joined, each tries to swing his body under the other and bring his hind feet and claws into action, driving terrifically against the belly of the other cat. Clinching, rolling, tearing, squalling, and screaming, they tumble over and over, first one on top and then the other, but the one who does the most belly-scratching wins. At last one cat will give a scream of pain, tear himself loose, and run away—often to die, because belly wounds are usually fatal.

The male cat's song to the female is another form of caterwauling, more like a wailing tremolo, and a piercing sound most people don't appreciate at midnight! The female may answer in a higher-pitched tone, but usually she is quiet, approaching her prospective mate through the shadows to look him over.

Most of us have heard the mewing of kittens calling their mother to come and bring them milk. That sound turns into an agonized wail if she is gone too long. But after the milk is in their tummies, they curl up in great luxury and purr themselves quickly to sleep. Both the young and the adults give a squall for help if they are caught by an enemy. This turns into an ear-piercing scream if they are hurt.

They have different ways of talking to men, and there are many cat personalities, largely developed by the attitudes and habits of the people they live with. Cats learn quickly how to get the best from humans. They learn whom they can bully and whom they must wheedle or flatter to get what they want.

Some spoiled darlings have chairs they consider their own, and anyone who disturbs them gets first an annoyed "Meow!" and then a blow with sharp claws. A cat of the other extreme, who has been treated roughly by its master, mews with fear and runs away when anybody comes near it.

One common way a cat communicates with humans is to rub

against a leg or to pull gently at pants or skirt with sharp claws. This is usually a request for food or attention, though sometimes it is a method of getting people to play. One cat I knew would rub and purr loudly against the legs of its master until he finally gave in and started off to get it a bowl of milk. Then, knowing the treat was coming, this pet would dash around his friend's feet, purring and rubbing against him in a paroxysm of delight.

These animals show they enjoy petting by lifting their heads and necks toward the petting hand and arching their backs. At such a time, a truly blissful look may come into a cat's eyes. However, I have seen them change in less than a second, suddenly turning and biting or scratching the hand that petted them. Possibly some part of the body was "rubbed the wrong way"; or they may have simply become bored with being petted too long. Sometimes cats are very unpredictable.

Kittens, and to a lesser extent adult cats, are great explorers, and are often very curious about new objects. A cat or kitten expresses curiosity by a widening of the eyes, an alert cocking of the ears, and a stealthy approach, almost as if it were hunting. When it finds a new object, it sniffs it all over and proceeds to test it with both teeth and claws, though very tentatively at first, by reaching out and touching the new thing timidly with the tip of a paw.

Some cats scratch at a door to be let out, but most usually stand in front of it and meow forlornly. One funny cat character who was anxious to get outside went first to the front door and discovered it was snowing, then to the back door, and finally to each window, hoping the snow would be gone!

Cats are often great admirers of themselves, and some will sit in front of a mirror for hours primping and cleaning. A cat will, in particular, clean itself with paw and tongue after a dangerous experience or when it has fought with another cat. This is not so much to make itself look good after being mussed up as it is an attempt to soothe the nerves by doing something natural and instinctive.

HORSES

In some ways horses are not as intelligent as dogs and cats, but in their expressions of affection and love toward a human master, they may outdo even dogs. A true love between a master and a horse brings the horse running at a whistled or shouted call, not just for the apple or piece of sugar that it may know is waiting, but also for the pure joy of comradeship. More than any other animal, such a horse loves to run its soft nose over a hand, arm, and neck.

My wife, who was raised on a cattle ranch in Colorado, has tamed and ridden numerous horses. I have seen her ride her favorite horse, Laddie, bareback at full speed with nothing but a single piece of string as a bridle. She signaled him to turn left or right with a pressure of her knees and a slight pull on the string. In such cases the feeling of love and cooperation passes from the hands and the legs into the horse's body, and back again like an electric current, and horse and rider seem to understand each other perfectly.

When I rode Laddie, there was no such recognition. My control over him depended on my strength with the bridle. When he wanted to run, there was no stopping him unless I exerted all my strength to pull back on his head. He showed his resentment for this unwanted signal by throwing his head from side to side and crawfishing sideways in full gallop so that I was hard put to it to stay in the saddle. When I brought him to an abrupt halt, he would sometimes rear to show his indignation. Clearly horses sense the difference between riders.

A big white horse, Eagle, that I owned in Panama, could be captured only by being fed stalks of sugar cane, of which he was very fond. Once haltered, he showed his resentment by trying to bite me, pulling back his ears and baring his big yellow teeth. This gradually changed as I lost my early fear of him and became more firm. A horse always senses when the rider is a little afraid of it, and sometimes seems to test a stranger by biting, bucking, or kicking out.

Horses meeting for the first time, or after an absence, sniff each other's noses in a kind of test greeting. If they are friends, they are soon either nuzzling each other in a friendly way or grazing together. But if they are former enemies or one horse doesn't like the smell of the other, back go the ears in the symbol of anger and one or both may jump at the other, biting and kicking. At the same time, they snort and whinny with anger.

Wild stallions attempt to gather harems of mares. They may round up a mare who tries to run away, but stallions do not lead the herd. This is always done by the most experienced mare. She signals to the herd to wait by stamping her foot and staring ahead. When she is sure

The horse obeys the good rider on the left, but acts up in a show of independence with the poor rider on the right.

it is safe to go on, she starts forward again, the herd following her. If the leader turns and runs, the herd turns with her as one horse and rushes away, for each has been watching her carefully.

If a strange stallion or gelding approaches the herd, however, it is the stallion who takes command. With many nips at their buttocks, he drives the mares away in a tight bunch. Then he gallops toward the stranger, snorting and whinnying with anger. His ears are laid back, his lips are drawn up to show his teeth. If the stranger does not run, he is attacked furiously with slashing teeth and hoofs.

Young colts talk to each other with the language of movement. It is all glorious play, a running to and fro at top speed. Up goes the tail of a colt in the signal that he wants to run. Up go the tails of all the other colts that see him, and away the bunch goes, recklessly, sometimes whirling around and kicking at each other in sheer exuberance of spirits. But if danger nears, the whicker of the colt calls his mother and she comes running with an anxious whinny. She hovers over him, her body guarding him, and the feel and touch of her is a message of comfort to him.

You can sometimes talk to a horse by breathing in its nostrils and speaking to it calmly in the highest voice you can manage. Some wild or skittish horses have been remarkably tamed in this way. Always approach very slowly and without the slightest fear. Stand still when near, and let the horse approach you, if possible. Be very patient—and finally bend over so your nose comes close to its nose and you breathe right into the nostrils. Do not attempt to pet the horse until you are sure it is friendly, and then move your arms very slowly.

The trembling of a horse's body shows you that it is afraid. Soothe it with a gentle voice and petting. A horse that is tired droops its head nearly to the ground. A horse that is thirsty blows flecks of foam from its lips and whinnies and snorts when it smells water, then rushes ahead at full gallop, if you let it. When it smells or hears a distant horse, it whinnies loudly; this is both a greeting and a desire to see the other, a desire that is especially strong if the horse has not seen another for some time. Horses love the company of their kind.

Some maverick horses may be quite dangerous. If you see a horse that is biting savagely at other horses, it is better to stay clear of it

Two stallions fight over mares.

and have a stick ready. You can usually tell this kind of maverick horse by its red-eyed look; it is warning you not to come near and may even attack you. There is a feeling that comes to one from such a horse, which some people can sense and some cannot. Keep your eyes open and be alert for that feeling. If a horse communicates with you in that way, it should be strictly left alone or driven off with a stout stick if it tries to attack.

We have three horses now: the mother, Epomo, her big gelding son, Nomad, and her petite daughter, Katy-bar. They dominate each other in that order. Nomad has only to bare his teeth to make Katy-bar leave some hay that he wants, but Epomo is the boss over Nomad. Sometimes one of these two has tried to boss or intimidate me. They don't any more, since I no longer show fear of them and when they act a bit nasty, talk to them sharply.

Once I had a remarkable experience with the three horses. It was nighttime, with a little moonlight. I heard them snorting down in the far pasture with a sound I knew meant something bad was up! Then they came running pell-mell for the barn and corral. I got the gate open for them just in time, for they rushed madly into the corral at

A mare controls her offspring through attitude and voice.

full gallop, swirled around in a circle and threw their heads as high as they could, looking over the corral fence back to where they had been. I could see they were deeply afraid and was sure it must be either a big bear or a mountain lion that had scared them. I knew I had to do something to calm them and make them respect me as their master. I grabbed a flashlight and a big stick, and started down the way they had come. I had to screw up all my courage, as their fear was infectious and I could not be sure what I would run into.

When I got down to the far pasture, my light caught up with what they had seen. It was a bright-red and brown Hereford steer that had somehow wandered over the mountain to the north from a ranch about a half mile away. I laughed with relief and drove the steer away by waving my stick. To the horses, I was the hero who had saved them from a monster they had never seen before.

CATTLE

Anybody who has been chased by a savage bull, and lived to tell about it, knows the warning signals—or should! I was once chased through eight-foot jungle grass in Panama by a big brindled bull and got over a fence just in time. I heard, first, a pawing on the ground, then a loud snort, and last the crash of a big body plunging through the grass. I have been told by those who should know that in the last resort you should throw your body on the ground and lie perfectly still. The bull may nose you a bit but otherwise leave you alone. It is the moving object that he is interested in goring.

Most bulls are comparatively peaceful if you do not tease them, especially range bulls (like the Herefords), who have plenty of cows to interest them and lots of room. But when you see a bull encircled by the fence of a small field and all alone, look out!

In the fights I have seen between range bulls, the first signal of hostile intentions is the throwing of dust. The front feet churn up the ground and throw as big a cloud of dust as possible over their backs. This is sometimes done also when they are fighting off flies, but when two bulls face each other, it usually means war.

They soon begin to snort and bellow their rage, working themselves up to make a charge. They get nearer and finally one bull gives a bigger bellow and rushes at the other. The two meet with a smacking noise of horns and skulls, shake their heads to clear away the shock, back off a way, and charge again. If one bull can gore the other in the side or force him back by superior strength, usually the other bull turns and runs, and the fight is won.

Cattle are very curious. When one of a herd sights something strange, it turns and faces toward the object or person with an open mouth, the ears spread wide and the big nostrils flaring as they sniff. In a few seconds every member of the herd has caught the signal from the first animal, and all turn and look the same way. If one cow or steer becomes alarmed, it throws its tail suddenly straight into the air (a warning signal) and turns and gallops clumsily away. Usually the rest of the herd does the same. However, I have known them suddenly to come rushing back from another direction, overcome by their curiosity to see what I am or perhaps hoping I have food or salt for them. Where cattle are half-wild and familiar only with men on horseback, they may actually charge a man on foot, snorting and bellowing to work themselves up into a rage.

The cow comes into periods every month or so when she is ready to mate with a bull. Our cow, Star, begins to bawl loudly on such a day and may keep it up off and on for as long as two or three days. At such times, she is also much more restless and more impatient about being milked. The bull either smells or hears the cow, or both, at such times and begins to bellow an answer as he approaches. Some cows are quiet and only give off the mating scent at this time.

Since Star has had four calves since she has been with us, we have had good opportunities to watch the communication between cow and calf. When the calf is very young, Star is much more anxious about it. She bellows a warning when she sees our dogs, and the calf comes trembling close to her. If the dogs come too near, she bellows again and rushes at them, swinging her horns. When I move the calf away from her, she rushes me, too, but stops about a foot short of me and moos, as if to say, "You be careful; don't you hurt my baby—or I'll have to horn even you!" The calf stretches his neck, opens wide his mouth, and lets out a loud "Maaa!" when he wants her for protection or is hungry. Her tongue, as she licks the calf, speaks a language of endearment, and she gives a soft moaning sound, which I call her mother-love song.

Cows have different ways of communicating with their masters. When the udders are too full of milk and hurting, the cow gives a low bellow of pain that means she wants to be milked soon. When bothered badly by flies, she constantly throws her head back around her shoulders, throwing spittle at the flies or trying to swish them off with a mouthful of hay. This means she is in need of spraying or powdering to get rid of these pests. When she is determined to get to her food in a hurry and wants to show her independence of her master, she shakes her head violently and rushes full tilt for the feed bin. When she wants water, she stands by the watering trough and bellows.

A cow we had later, named Daisy Bell, was even more independent. One day she broke out of her enclosure and went wandering over the hills, with me chasing after her. She kept running me until I was pretty tired—and pretty angry, as she would not respond to my calls. Finally I cornered her where two fences came together, and this time when I approached, she lowered her head and horns, pawed the ground, and bellowed just like a bull. She gave me the strong indication that she meant to fight if I came near her. However, I knew her well enough to know this was bluff. (Bulls should not be treated this lightly.) I came up to her as she was pawing all that dust into the air and said to her in a loud voice, "Daisy Bell, you stop that right this minute!" She calmed down immediately, stopped her pawing, and meekly allowed me to put a rope around her neck and lead her back to the fenced-in area. She showed she knew who was master, but we did have to make that fence stronger right away so she could not get out again.

SHEEP

Sheep are supposed to be very stupid, but we have raised sheep for years, and my wife did for many years before that. We find much that is of interest about them. They are considerably smarter than one would think about sneaking in through the gate and getting into flower and vegetable gardens. One old ewe hides near our gate sometimes and sneaks in when it is opened to let in a car. If the rest of the flock are near, they come right along with her.

As with most animals, the ewe's first job when the lamb is born is to lick it all over thoroughly. While doing this, she makes a peculiar little "Muh!" note, repeated over and over. There is a good deal of anxiety and love in this sound, as if the mother were saying, "Are you alive? Are you all right? Are you *really* alive?" Once the mother has finished the licking job, she has completely absorbed the special smell of her own young one and henceforth will never mistake her lamb for another. If twins are born, sometimes one twin is not licked in this manner; the mother rejects it and butts it away just as if it were a real stranger. Such a forlorn little orphan, who constantly wails in a treble voice for a mother who never answers, has to be raised by bottle, and is called a "bummer." In a day or two it decides that I am its mother and follows me everywhere!

When the normal new lamb is hungry, it bleats very softly for food. It butts around blindly with its head until, by accident, it comes in contact with one of its mother's nipples. Once it begins to suck, a blissful look comes over its face, and it closes its eyes. The sign of high happiness is its tail, which wiggles at a high rate of speed. Meanwhile the ewe, giving her repeated mothering cry, turns around and gently licks the rear end of her newborn. By this little drama of mother and baby, the two become united, each signifying its love and need for the other. The act is mirrored in every mammal family at the time of each new birth.

A ewe accepts her lamb.

A ewe nudges a new lamb to keep it active.

If you watch the tail of a sheep, you can tell a good deal about its emotions. When my sheep gather around me, excited over the prospect of grain, their tails are wagging vigorously. Then the tails sway slowly as the sheep contentedly feed. On the other hand, when a ewe wants to tell a ram she doesn't want him to bother her, she also excitedly wags her tail before making a quick run away from him. Should the tail be held down close to the body, it would mean the sheep is frightened or disturbed.

Two rams preparing to fight do so with much less fanfare than two bulls. They usually simply back away from each other with heads lowered, then rush together with a crashing impact of skull against skull. This butting of heads may last only a few minutes or go on for an hour or more, depending on the relative strengths of the two. Sometimes one or both may have their skulls seriously injured by the blows. More often, one ram, feeling he is the weaker, suddenly turns and runs.

You have to be careful with rams, particularly one raised alone by humans, as some are not afraid of people. We had such a ram, and he was completely untrustworthy. I had to watch him all the time. My daughters were taking some other girls through the pasture once, all laughing and chattering merrily, when suddenly the ram smashed into them! I heard them screaming and shrieking and rushed to save them. I threw the ram on his side and held him down until they could get to safety. Fortunately, no one was hurt badly, but feelings were very ruffled! You must know your ram.

A later and bigger "bummer," but one I thought was trustworthy, charged me from behind one day, coming so fast and silently that I was knocked flat on my face before I knew what was happening. I didn't retaliate then, but a few days later when he charged face to face, I had a boy's plastic baseball bat in my hand and broke that to pieces over his head. It did not really hurt him, but it so startled him that he has been peaceful ever since, though I still keep my eyes open! That time I could see that he was going to charge by the look in his eyes and the way he lifted and then lowered his head. That was the signal of the attack. If he had been successful, he would have been more dangerous later.

When a lamb is young, the mother sheep is much braver toward dogs and coyotes than later, and may actually lower her head to rush at them and butt them. Her feet are often stamped in warning and her ears laid back. At other times, when a killer dog gets among sheep, it finds them easy prey, for they all run, without any attempt to defend themselves. After one of our sheep was killed by such a dog, the rest of the flock were nervous for days and seemed to run from every shadow that moved. When running in fear, they tuck their tails low and give a loud "Baa-a!" full of fear.

The baa of a sheep can be varied to express various meanings. The lamb that calls its mother for milk or that is lost produces a high-pitched tone. An eager and loud sound comes from all the sheep when they think I am bringing them food. The ram makes a kind of guttural "Uh-uh-uh-uh" sound when he is wooing a ewe, and places his head close to her side. Then there is the basic conversational baa, in ordinary tones, of members of the flock keeping in touch with each other while feeding.

RABBITS

Most people think that all rabbits are alike, but we have found many very different characters among the hundreds we have raised. Our New Zealand white rabbit, Alice, always taught her babies to be very clean and never to put their droppings or water into the food and water dishes; Alice simply nipped their little behinds when they did something naughty, and they soon learned to behave. But many other mothers were quite slovenly.

When a mother rabbit is soon to have young ones, she begins to pull fur from her chest, neck, and body, and place bunches of it in the nesting box. She is extremely absorbed by and fussy about this job, as if the fate of the world hinged on her doing it just right. The babies are born a day or two afterward and are kept covered by the soft fur if the nights are cold, or left exposed if it is hot. They are naked and blind but capable of making tiny mewing noises when they are hungry or disturbed.

If a dog comes near, the mother hops about anxiously and stamps her feet; she would kick him if he came too close. But faced with a gopher snake, she reacts by crouching in paralyzed fear in one corner of her cage, allowing the snake to eat her young ones without protest.

Young rabbits, when they get a little older, squeal with great vigor if they are picked up. No doubt this cry is used to bring the mother to help them when there is real danger. The adults also squeal loudly when hurt or chased by a dog. This loud squeal may be of use in temporarily upsetting an enemy so the rabbit can kick its way free and run.

Grown female rabbits usually have a great dislike for each other because of the instinctive desire of each female to control a certain feeding territory and not allow any other female near. When put together, they first spar vigorously like boxers with their front feet, then try to secure a bite and curve the body over, kicking powerfully with the hind feet. Blows from these feet can actually disembowel another rabbit. The ears are laid back flat on the head at the start of these fights, and often the two mumble angrily at each other. Males fight in a similar way, but may do more boxing with their feet.

When a male and female are put together, the male immediately begins to nuzzle the female and nibble at her with his mouth, which is his way of wooing her. If the female is not interested, she hops rapidly about, mumbles and whines in an angry fashion, and violently shakes her head. She may even turn on him and bite or box, but he does not fight back. This angry mumbling whine is also used by one female to show her jealousy of another female who is with the male. Such a female expresses great annoyance by jumping about and throwing her head and shoulders from side to side.

The noses of rabbits are highly expressive of their curiosity and interest in each other and in other animals and things.

When a rabbit sees or sniffs something strange, it may squat back on its hind legs or even stand up to full height and wriggle its nose vigorously, trying to catch the scent of the strange thing. If it tastes something that it doesn't like, it sniffs and snorts violently, rubbing its nose quickly with its forefeet.

When it is curious, its ears cock forward and spread wide to catch the slightest sound. A rabbit expresses its health and well-being by sitting up and grooming itself with its front forepaws, first running the paw through the mouth to wet it and then combing the fur on its sides and head. A healthy rabbit likes to be clean, and a sure sign that a rabbit is not feeling well is when it begins to allow itself to get dirty.

Some rabbits have much stronger characters than others. We had five half-grown rabbits escape from a cage one day because I forgot to lock it. Four of them were fairly easy to run down, but the fifth and most beautiful one remained at large. My two grandsons, their mother, my wife, and I finally formed a large ring around this rabbit, and it did an astonishing thing. It jumped and ran and hopped in circles and crisscrossed the surrounded area, often at high speed, enjoying itself tremendously. When it finished, it simply ran over to me and by its look and actions showed that it wanted me to pick it up and take it back home. It had been actually showing off its powers to us. We call her Little Grey-Brown, and she is still very tame and likes to be petted.

CHICKENS

Chickens are among the most familiar of all domestic birds. My wife and I have raised various kinds and have had many fine friends among them. There was Turkmom, the hen who raised many baby turkeys; Carol, who could sing most beautifully; and Caesar, as gentle and friendly a rooster as ever lived. Though chickens vary in their personalities, they are still subject to certain rigid social rules that govern their relations with one another and how they communicate.

Baby chicks have a constant, shrill "Peep, peep, peep!" when they are hungry or cold. This changes into a slower, gentler, and more satisfied peeping when they are warmed and fed. As they get full and drowsy under the warmth of the brooder light or their mother's body, they each tuck a head under a wing and go to sleep. They are born with the instinct to peck, and at first they peck at everything and pick it up, but soon they discover that certain things are good to eat, especially the yellow grain, and after that they peck at anything yellow.

Before the chicks are many days old, little fights start among them. These are hardly more than play at first, but gradually the chicks learn who is stronger and, what is sometimes more important, who has the most courage. Gradually the flock begins forming what is called the "peck order." The bravest and strongest chicken is at the top of this order. When he pecks any of the other chickens, they run away from him. The second bravest and strongest chicken can peck all the other chickens except Number One. So it goes, until finally one poor, lonely, little chicken can be pecked by everybody else, but has not one it can peck!

When one rooster meets another rooster before the peck order has been established, they both spread their neck feathers into a ruff and stand as high and stiff as they can on their legs. This is the challenge to fight, and, if neither backs down, they fly at each other, pecking with bills and striking with the spurs on their feet. If nearly even, the fight may go on for a while and even end in a draw. Otherwise the weaker rooster suddenly lowers his tail feathers in a sign of defeat, ducks his head, and runs for his life. After that he just runs whenever the other rooster approaches.

Hens appear to be able to spot something nice to eat and rush for it long before the roosters catch on. Every chicken in sight almost immediately sees the running hen and runs, too, to see what it is she is after. Thus the movement acts as a signal.

There are several calls given by chickens, hens having more of them than roosters because of raising the chicks. The loud cackling of a hen who has laid an egg expresses deep pride and is usually answered by a rooster who makes a loud cry of joy and announcement, as if he had done the work as much as she had! This is always a big moment in the chicken yard. Another call is the loud, squawking cry of alarm when a hawk or other large bird is sighted. This sends most of the chickens scurrying for shelter. The cry is usually given by an old hen, wiser than the rest.

The crow of the rooster, in the morning or at other times of the day, is simply an expression of his strength and maleness, the rooster

An illustration of the "peck order" among chickens.

who is king of the chicken yard giving the deepest and longest crow. We had one young rooster who crowed so badly that the other rooster, our beloved Caesar, would give him a disgusted look, sit back on his heels, puff out his chest, open his beak wide, and give a thorough demonstration of the correct way to crow!

Whenever we put fresh straw in our chicken nests, a half dozen or so hens jump up on the roosts to oversee the job, every one of them crooning a little song of delight over getting a clean nest. This crooning is a signal of pure happiness. A somewhat similar trilling is heard at night when the chickens settle themselves down—warm and comfortable on the roosts, feeling protected, among friends, and ready for sleep.

Suppose one chicken is feeding in the feed trough and another chicken of lower peck order comes up too near the same place to get a bite. The first chicken feels insulted, ruffles up her feathers in warning, and turns and makes a sharp peck at the intruder. The answer is a shrill squawk of outrage and hurt, and the lesser chicken stalks angrily away, also ruffling all her feathers. A somewhat different message is given by a rooster who has found some overlooked grain or a nice, luscious worm. His idea, instead of keeping all the food to himself, is to call all the hens over so he can show them what a fine fellow he is. He immediately starts up a rapid "Cuck-cuck-cuctckt!"—meanwhile vigorously scratching the ground. At his call, all the nearby hens come running; he proudly watches them eat, strutting up and down to show off his plumage.

The hen with baby chicks has a somewhat similar call when she finds feed for her young ones. "Cut-cut-cut-cut!" she calls rapidly, and all the babies come running up to where she is vigorously scratching the ground and pecking. About half the time she drops the picked-up food from her mouth so the babies can get the idea and eat it. Another, softer cry from the mother, a kind of soothing "Ptt-ptt-ptt," brings the little ones under her widespread wings when a rain is coming.

A broody hen, sitting on her eggs, may give sleepy, contented noises, very soft when she feels protected; but let someone come to take eggs out from under her, and she will often fight desperately to save them, even drawing blood. However, I have found that by talking gently to the hen in this situation, I can often get away with a few eggs without a fight. Scientists have described nineteen different calls made by chickens, which may indicate more intelligence than we attribute to them. See if you can hear other calls. The terrible screaming squawk of a hen that is being dragged away by a fox is one, and another is the loud shriek, given when a hawk is noticed overhead, that sends all the chickens running for cover. In the former case, the loud squawk may actually make a fox drop the bird.

Our Airedale, Jill, came around a corner one day and ran unexpectedly into a mother hen and her chicks. Instantly the hen's feathers puffed out in all directions, making her look twice her normal size, and, with a loud war cry, or squawk, she leaped right at Jill's face, striking with both beak and claws. Jill let out a yelp of fear, whirled and ran, with the triumphant mother hen practically riding her back and digging in with every toenail! Meanwhile the chicks had obeyed the ancient call to scatter and had run in every direction to hide in the grass and bushes. When the mother dropped off her fleeing enemy's back, she came back and began a series of soft clucks that gradually pulled her brood together again.

The rooster who is wooing a hen does so not only with his loud crows and his scratching of the ground, but often spreads the feathers of one wing so they scrape the ground, and circles around her with a sidling, scuffling movement of his feet. If she doesn't like him, she runs away; but if she does, she crouches to the ground. Some hens show their love for a rooster by gently pecking at his bill.

TURKEYS

We have raised hundreds of turkeys and find them in some ways more delightful than chickens, though often less intelligent. Young turkeys, unlike young chickens, have to be taught both to pick up food and to drink water, as they seem to be utterly helpless when just out of the eggs. However, when put in with baby chicks, they soon copy their techniques.

On the other hand, turkeys do something in the early dawn that seems to me quite wonderful. They all stand facing the east and appear to watch in silent ecstasy as the sun starts to rise over the hills. It is a moving sight!

Domestic Animals

One thing I like about a young turkey is its joy at being alive. When we let them out of their night pen into the wide and golden freedom of the turkey yard, they spread their wings and flap them gloriously, and then rush about flapping and jumping high, at the same time uttering a high-pitched "Peep-peep-peep!" that expresses unlimited delight.

The peck order is not nearly so rigid among turkeys as it is among chickens. If one turkey "gets a mad on," it seems to be able to rout even a bigger turkey, provided the other doesn't get just as mad. A fight between two turkeys is quite different from a fight between chickens. The first sign of trouble coming is when one or more turkeys start to spread their tail feathers and wings, stretch out their necks, and begin to call names. One turkey sidles up to another turkey and gives a sharp peck. Immediately the turkey war cry increases in volume!

This cry, a high-pitched "Krooot-krut-krooo!" is repeated rapidly over and over as long as the fight lasts, and expresses both bad temper and an attempt on each side to bluff the other into quitting. The expression on the faces of both turkeys in the fight is strictly nasty, almost ludicrously so. They are so intent on doing each other bodily harm that nothing else matters. Back and forth they strike and peck until one gets a deathlike grip on the other's wattle. Then the caught one twists its neck in all kinds of snakelike contortions, trying to break free.

Turkey fights usually stop when one of the turkeys breaks loose and makes a run for it. However, as the fall mating season approaches, battles become more and more serious. If blood is drawn, other turkeys may join in to attack the bloody one, and before long this weaker bird is pecked and trampled to death by the mob. Sometimes a turkey can save itself by stretching its neck out along the ground. This is a signal of submission and usually ends the fight.

Tame turkeys have smaller brains than wild turkeys and are not nearly so well equipped to protect themselves from danger. However, they do give each other warning. At night I have heard their "Prrrrt-prrrrt-prrrrt!" exclamation of alarm and crept out to see what was wrong. Twice, because of this warning, I have caught raccoons robbing the chicken house. The turkeys on their high roosts in the turkey yard had seen a strange, dark animal sneaking by the fence and had

Turkeys surround a snake and prepare to kill it.

given the old cry of alarm to the flock. In the daytime, I have heard this same cry announcing a snake in their yard. When I arrived, I would invariably find them standing around the snake in a half circle, their necks craned toward it, uttering their warning, and gradually working up their courage to attack and kill it.

When the mating season draws near, the big tom turkeys begin to strut and spread their feathers before the hens. Their wattles get very red; they tuck their chins proudly back into their throats and move forward in jerks, at each jerk making a noise deep in their throats that reminds me of a car's motor starting. The females watch them admiringly, and when a hen wishes to show she is deeply interested in a particular tom, she walks up and crouches on the ground in front of him, making a low, gentle "Qt-qt-qt" noise.

When danger approaches from the sky in the form of a hawk, turkeys crane their necks and look upward. If there are baby turkeys about, a hen suddenly gives a loud "Prrrrrt!" cry and all the babies run for the shelter of the bushes, squatting down and making themselves as inconspicuous as possible. At such times they hold perfectly still, without a sound, and are very difficult to find.

Turkeys we raised under the brooder came to regard my wife and me as their mamas and would follow us around, giving the high "cheep" of hunger until we fed and watered them.

The gobble of the full-grown tom is an expression of pride and is probably also a way of showing off to the female, but it may also be an answer to a challenge. When I shout at a flock of turkeys, the toms gobble back at me almost as one bird, stretching out their necks and shaking their bright red wattles. Possibly this loud noise from a flock has some value in frightening away a potential enemy.

DUCKS

Ducks are the clowns of our barnyard. They do the most amazing things! For instance, one female duck followed the rooster all over the yard, curving her neck coyly at him and constantly quacking until the poor rooster didn't know what to do! The quack of the duck can go through several shades of meaning simply by its change in tone. When I try to catch a duck, she quacks so loudly and with such great fear in the tone that it sounds remarkably like a terrified woman screaming, "Help, help!"

When we have cleaned out the pool for the ducks and they approach the fresh water, they act as if they cannot believe their eyes—they are so delighted! Luxuriously they stretch their necks and heads down into the water, and sip it up with a look of utter rapture. Then one of them runs in a crazy circle, quacking with low-toned joy, and suddenly wheels back and dives into the water. Up and down under the water it goes and then comes to the surface, beating it with wings

Ducks love to play in fresh water.

and giving a half yell, half quack of triumph that is a call to the other ducks to come join the fun.

Ducks seem to like to talk to each other socially much more than chickens do, for their constant quack-quack of inquiry and friendliness is heard repeatedly, especially when they can find some shallow water in which to run their bills. This bill-running through water—in and out, opening and shutting at high speed, probably taking in tiny things to eat—is in itself a kind of conversation. Two ducks will run their bills through the water almost side by side and often touch bill to bill as if kissing, all the time seeming to enjoy themselves greatly.

In the mating season, the male duck follows the female into the water and talks to her by constantly preening himself, showing his curled tail feathers, and ducking his head up and down with a peculiar motion. They quack to each other, often at high speed, and soon she begins to duck her head up and down, too. When you see the two rubbing their long necks together, you know the male has made a conquest.

CANARIES AND PARAKEETS

Canaries have long been the most popular of indoor bird pets, but their place is being challenged these days by parakeets, or "budgies." Parakeets are more intelligent than canaries and have a more extensive language.

We kept canaries for fifteen years and loved the clear beautiful singing of the genuine German roller canary. The song is a pure expression of joy at being alive, combined, in the mating season, with the longing of the male for his mate and a challenge to all other males to stay away from his territory. The most intelligent of our birds was named Peppy, and he loved to fight with a human finger, attacking furiously with his bill and cheeping a savage war cry.

Male and female canaries express their joy in one another by rubbing bills together and running each other's feathers through their beaks. Each takes turns feeding regurgitated food to the young ones, who signal when they are hungry with wide-open bills and loud cheeps.

The ordinary "Peep? Peep?" of a canary, given with a rising note of inquiry, is usually just a social call from one to another, but sometimes it means a bird wants somebody to pay attention to him. Our birds greatly enjoyed conversation, and would cheep back a reply, as if to say, "Tell me more!" When we freed them from their cages, they would fly round and round the room and often light on our shoulders with a cheep of triumph at their own daring.

Parakeets, with their greater intelligence, are up to many more tricks. Their light twittering cry can be varied in tone to give different emotional meanings, including one that seems to me very close to laughter, when they have done something naughty and think they are getting away with it. They can be trained to do many clever tricks, such as taking food out of the lips of their master and walking upside down along a thin stick. The bill of a parakeet acts as a third hand for climbing. The bird expresses emotions of anger or nervousness by snapping its beak, and hunger by opening it wide and calling plaintively.

The parakeet male and female love to preen each other with their bills, going carefully over each feather and crooning softly all the time in a kind of love song. They also pass regurgitated food back and forth, and this seems to be something in the way of a kiss, for they greatly enjoy it. But they can get into loud squabbles very easily if one does something the other doesn't like. Probably a whole book could be written about the language of parakeets and parrots.

4
The Meat Eaters

Meat-eating animals occasionally eat one another. Thus the hunter becomes the hunted, and the actions and language of the animal who is hunted greatly change in tone and meaning. The large weasellike animal, the fisher, may chase down and eat the smaller member of the weasel family, the marten. When the marten finds itself chased by its large cousin, it reacts with the same fear actions and cries that drive the squirrel to terror-stricken flight before the marten.

Carnivores (the meat eaters) are generally of more fierce temperament than the herbivores (the plant eaters); the structure and muscles of their faces are adapted for more vivid display of rage and threat. The wildcat, crouching over a mouse it has just caught, pulls back its lips ferociously from its long, sharp canine teeth in a snarl of warning. Its tail lashes furiously back and forth, clearly saying, "Stay away, this is my food!"

THE WEASEL FAMILY

Members of this family, with the exception of the skunks, are all strictly meat eaters, and consequently all their actions are based on the chase of prey. With the exception of the badger and the wolverine, their long, slim bodies are particularly well-adapted for following animals into holes.

We may divide the weasels into three general groups:

1. the solitary, instinctive chasers and killers, including most of the true weasels;
2. the intelligent and semisocial otters;
3. the amiable, omnivorous, and lazy skunks.

True Weasels

Most weasels, mink, badgers, martens, fishers, and wolverines are solitary creatures who come together only for brief periods at the mating season. They have little need for communication and so their language is rather primitive.

I once watched two long-tailed weasels in the Sierras hunting for mice in a pile of logs. They may have been mates, but there was no cooperation in their hunting. Each was silently intent on its own business, and the chief language they used was one of movement. A twitching of the nose and a swinging of the head in an arc from side to side told me they were trying to catch the scent of a mouse. A lowering of the head and a sudden flash of movement as the weasel dived down between some logs told me it had caught the scent and was on a hot trail. A shrill squeak from a mouse told of terror under the logs, but the weasel was silent.

One weasel leaped up onto a log with a mouse in its mouth and proceeded to tear its prey to bloody shreds and gulp them down. When the other weasel passed near by, the first one arched its back, opened its mouth, drew back its lips, and made an almost soundless

A weasel snarls to indicate possession of the wood rat it has just killed.

hiss. This was sufficient warning to stay away, for the other did not come near, but continued its own hunting.

The almost snakelike hiss of warning seems to be common to most of the weasel tribe, including the otter. They also have a harsh snarl, deeper in tone in the larger animals, and a blood-chilling scream that is fortunately rarely given. The snarl is given in bluffing and also in actual combat, and it is always a sign of mean temper or rage. The scream is a sign the weasel is in mortal agony, deadly fear, or insurmountable rage, as when its food is stolen from it.

Young weasels, martens, and their like produce a cry that is a kind of cross between a mew and a whine to call their mothers when hungry or lost. Sharp snarls from the mother warn them to be silent or to hide. I once heard young weasels playing in a pile of brush. They were chirping softly to each other almost like birds, moving with such speed in an apparent game of tag that only flashes of their bodies were seen.

True weasels seem to move and act mainly by instinct, showing little of the curiosity that goes with intelligence and sensitivity. The marten, the fisher, and the mink, however, show some curiosity. I have heard of a mink carefully exploring a dam; and a marten once tried to investigate me, until it discovered I was alive!

Young weasels sometimes give a curious, crowing note when their play is the most fun. Very young weasels squeak hungrily for food. The mother short-tailed weasel has been heard to call her young with a kind of grumbling "Coo-oo" that seems to mean "Come on, children, we must be going home now." A shrill snarl or scream of warning sends them tumbling down a hole into hiding, or even makes them seize their mother's fur tightly in their teeth so she can carry them away at full speed.

The mother and, more rarely, the father will set examples for the young, which the young copy, at first crudely and then with more and more ability as they grow older. I have seen a mother weasel in captivity drag a wounded mouse near her young and, by pretending to pounce, urge them to catch it. They quickly sprang upon it and tore it to pieces with delighted and shrill snarls.

In the mating season, mainly in late fall (though sometimes in the early spring), the attraction between the male and female is accompanied by the hatred between males. If two males meet and one is smaller, the larger male puffs up his body, sticks his hair straight out, arches his back, lifts his tail high, and advances menacingly to the attack, while giving a deep and savage growl of warning. The smaller male crouches low, hissing and spitting his rage, uncertainty, and fear. Usually, at the last possible moment, he turns and runs. If the two males are nearly equal in strength, both act as the first male described above, but they advance more slowly toward each other, sometimes slowly circling. One may successfully bluff the other, or each may have such deep respect for the other's fighting ability that a truce will be declared and they will back away slowly and go their separate ways. Often, however, there is a fight to the death.

A female, when she is ready for mating, may attract the male with low clucking sounds. She may run from him a little way, but her running is full of invitation in every twist of her sinuous body, and she often looks back. When he gets too close, she may snarl at him and even bite, but her snarl and bite are only half serious.

To the male, one of the main attractions of the female lies in the scents she leaves along the trails. Her scent glands are in the rear of the animal and give out a smell that is quite vile to men, and usually obnoxious to most other animals, but it is an aroma that seems to be delightful to weasels. The male also leaves his scent here and there in the woods, partly to attract the female and partly to warn away other males. Weasels can tell from these scent deposits much about another animal. A male tells us by his attitude of anger or of eager interest whether the scent was left by a rival male or a friendly female.

Mink. The mink, because it spends so much of its life in the water, has somewhat different reactions from those of other weasels. It loves to explore streams and ponds, and the snarling screech the mother gives when there is danger may cause the young to dive into a dark pool as willingly as into a hole in the ground. When hurt or excited, the mink may give a remarkable shrill, twittering squeak that is almost birdlike. The loud sniffing of one mink is a signal to another that it has found something interesting to investigate. When one mink finds the droppings of another, it stops to study them carefully, especially if it is in a new neighborhood. It may thus be able to tell what food supplies are available.

Badgers. The badger is probably the most solitary of all the weasel tribe with the exception of the wolverine. Since it hunts its food mainly by digging out ground squirrels and other burrowing rodents, much of its life is spent under the ground. The immensely powerful and heavily clawed front legs are specially meant for high-powered digging. I once watched a badger dig itself out of sight in the Nevada desert in less than a minute! When he has found a hole that contains a live inhabitant, he tells you so by the snuffling noise he makes as he sticks his nose deep into the ground.

My wife and I followed a badger for several hundred yards one day in the Great Plains of Colorado. This one made no attempt to dig to escape us, but scuttled along with its flat body low to the ground. When we got too close, it would back into a bush with its head facing us and snarl and hiss ferociously. Its hair stood on end, and it seemed to spread and flatten its body even more than usual in a clear warning not to come too close. This trick of retreating backward into a bush comes from a need for protection against dogs and coyotes, who are thus forced to face the formidable teeth and jaws. As we followed it, occasionally we would catch a strong and bad-smelling whiff from its rear-end glands, clearly let out to discourage our attack.

In the early part of the year, the male badger begins to look for a mate. He rubs his tail glands on bushes and trees, and leaves a smell to lure the female and at the same time warn away other males. A grunting noise from badgers at this time of year probably is something in the nature of a mating call, the male and female hailing each other. The female bites and snarls at first when approached, but finally the two blissfully rub noses and sides together.

Otters

Otters, both those of the sea and those of inland rivers and lakes, are among the most humanlike of all animals. The affectionate family lives and the social gatherings of relatives to play and talk are very much like those of men. Otters scream when wounded, hurt, or frightened, in very human tones, and their liquid, dark eyes are deeply expressive of pain and even pleading. Otters have even been known to kiss and to fondle one another, while making sounds of endearment.

The river otter is more solitary than the sea otter, which gathers together in herds, or pods, as they are called. Each male river otter has a special range that he guards from intruders. This range may cover a length of fifty miles or more of river and take over two weeks for the otter to cover thoroughly. To this range he lures his mate, and in it the two raise their young. Often the male and female mate for life, and they are very kind and hardworking parents, taking great pride in their children, whom they watch over fondly.

Many movements of the otter have meaning to its own kind, and to us, if we watch closely. A quickly lifted, twisting, snakelike movement of the head is used when the otter watches for danger. It may turn its head slowly, a brief swing at a time, sniffing deeply of the air, when it is trying to catch the scent of an enemy or another otter. An otter springing up out of the water of a river with a fat trout in its teeth certainly is a picture of the self-satisfied hunter. It shakes the water off its dark, sleek fur with a gesture of triumph. While an otter has not been known to slap the water with its tail, like a beaver, the sudden, loud splash of the father otter as he hits the water in a swift dive is much louder than the usual sound of his dive for fish, and this splash is clearly a warning to wife and children to get out of the neighborhood in a hurry. I have heard the warning twice on the Eel River in California.

On the same river, I once watched an otter swimming swiftly through shallow water, trying to escape me. The body twisted and turned among the rocks like a dark eel. When it reached a deep pool, it surfaced briefly, snorted loudly by blowing water from its nose, and dived into the dark depths under overhanging rocks. The snort was a warning to other otters that a deadly danger was near.

Both young and old river otters love to play endlessly on slick clay slides down the steep banks of rivers. I have watched them hit the slide and shoot down it, to end with a satisfying splash in the pool below, with every bit of the delighted expression of small boys sliding down a good snow hill on sleds. While playing, they talk together with low, birdlike chirps that have different meanings according to the inflection, such as "Hurry up—it's my turn next!" or "Boy, is this fun!"

A kind of satisfied grunt is given by the otter when it has eaten its fill of good food or is having a good sun bath. Another very different sound is its shrill half whistle, half squeal—very short and sharp and often with a peculiar quality of anger or hate. This is usually given when the otter feels its rights are being infringed upon, such as when another animal steals its food. If the otter feels it can beat the thief, it backs up this squealing whistle with a more savage growl or snarl. A sniffing whicker or low snort is given by an otter when inquiring of a companion about food or danger.

Occasionally, the cry of the otter rises into a hair-raising scream—extremely shrill, and one of the loudest and the farthest-reaching sounds of the wilderness. This may be its supreme cry of rage, hate, or defiance.

When the male river otter finds a friendly female, the two play and romp together like children, chirruping gently. The touch of nose to nose and side to side also conveys a message. Both mates may have first sent out signals by the smell stations they leave along a riverbank. These are piles of dung and urine mixed with mud, each with its distinctive individual smell that tells another otter much about the individual who left it.

The love of otter pairs is so deep that often, if one of the pair is killed, the living one becomes very despondent and chirrups sadly for its lost mate for hours on end, seeking and seeking for the one who is gone. He (or she) may even refuse to eat and actually appear to suffer from a broken heart.

River otters generally mate in late winter or early spring, and the young are born about two months later in April or May. The father is generally away during the early part of their life; when he joins the family, the two parents begin to teach the young all their wisdom of the wild. Young otters cannot swim without being urged by their parents, who chirrup encouragingly from the water. If they stubbornly refuse to come in, the mother may take them on her back, swim into deep water, and then slowly sink from under them.

Later the parents teach them how to sneak up on a fish, turn over stones for insects, and dig through the mud for eels. The mother urges them to these tasks with a peculiar mixed whine and growl that seems to say, "Try again and do better, children! You can, you know!"

At the first sign of danger, the mother gives a loud chirruping and clucking sound, almost like a mother hen, and leads her young to safety, while the father, by his snorting and splashing, may try to lure away the enemy in the opposite direction. If a strange otter comes near, there may be a good deal of hissing in warning, and, unless this scares it away, a fight.

Unlike most of the weasels, otters rarely are driven by the savage killing instinct of the smaller weasels, which may kill many more creatures than they can eat. The otter kills only when hungry and then rarely more than it needs for food. If sad or lonely, it may not eat for many hours or even days. But when happy, with its fellows or even alone, it is far more playful than most weasels—tumbling, rolling, playing tag and follow-the-leader. A lone otter has even been observed playing with a smooth stone by tossing it into the air and catching it in its teeth.

Sea otters play and fight, and hunt for mussels and other seafood among the vast kelp beds of the Pacific Ocean—where the mothers rock their children to sleep in the natural cradle of the waves—but they do not form families as land otters do. When men are sighted, loud cries of warning cause the otters to hide among the kelp; men, by ruthless hunting, have almost wiped these animals from the face of the sea.

Skunks

Long ago, in the morning of the world, one member of the weasel family learned that it could throw a bad smell from its rear end and so help defend itself by gas warfare. This discovery and its subsequent development and improvement in future generations produced a revolutionary new character in the animal world, the skunk. The skunk does not have the speed either to catch animals or to run from them. It cannot follow them into a small hole in the way a weasel can. So, for food, the skunk has to eat almost anything it runs into that it can overpower, including insects and worms and some plants, and it must use careful stalking or ambushing to catch such fast creatures as mice. But for defense against enemies, it is well prepared, so well that it usually has no fear of attack and is consequently quite lazy and even friendly in nature.

Because of the ability to shoot powerful jets of bad-smelling fluid and gas a distance of eight to twelve feet, the skunk seems to say when you meet it, "I'm minding my own business, and you just mind yours. I don't want a fight, but I am not afraid of you. You keep away from me and nothing will happen, but if you get too close, you'll wish you hadn't!"

I recently followed a striped skunk for a distance of more than a mile. At no time did it travel faster than a man could at a good fast walk, even though my two dogs as well as myself and my four-year-old daughter Tamara were constantly pressing near to watch. Occasionally it would stop and stamp its feet, which is a solid skunk warning that means, "You are getting too close!"

Once, when I came to within ten feet of where this skunk was ambling up a dry creek bed, it shot out two yellowish jets of liquid in a fine spray for a distance of about four feet behind. The powerful smell struck my nose like a battering ram, and I drew back very hastily!

Skunks vary greatly in character. I have met skunks who were very nervous and ready to throw their scent with little warning, and others who were quite hard to stir up and obviously so friendly that they would throw gas only under the most severe provocation. The usual skunk goes through three warning signals before spraying. First it turns toward you and stamps its feet, looking at you intently out of its small, black eyes. The second warning is a raising and spreading of the tail—all but the tip, which hangs downward. The third warning usually is given just before it lets loose with a salvo from its guns: it throws the tail tip up sharply and spreads out the hairs. That means "Jump, *quick!*" The little spotted skunk of the West sometimes varies these warnings by standing on the front paws only, like an acrobat balancing on his hands.

The smell of skunks is no doubt also used to give signals to other skunks, especially during the mating time in late winter. The skunk smell, coming through the woods on a damp day, is genuinely pleasant to a real outdoorsman, for it gives a sense of the strangeness of the wild. Each such scent conveys to other skunks a knowledge of the one who made the smell, whether it is a he or she, and whether it is in a mating mood or not. Unmated females often let off little explosions of musk along the trails, to tell possible boyfriends they are ready for company.

When male skunks approach each other to fight over a female, they give low snarls, the more scared animal snarling the louder. They may stamp feet at each other, all a part of the bluff to scare the other away. If a fight actually starts and one skunk feels he is losing, he is apt to squeal very loud and long, and then turn and run away. The snarl and stamping express anger and warning. The squeal is the hateful cry of one who realizes he is losing.

A spotted skunk prepares to throw its scent in self-defense.

Baby skunks are usually born in the spring and are generally looked after by the mother alone, though occasionally the father has been reported to help. The mother and children have several communicating sounds and movements. If the mother senses danger, she may turn toward it with her nose out and her whole body tense. Instantly all the little skunks turn and tense. If she wants her babies to stop their play, she may give a low churring, scolding cry that brings them running to her. A birdlike twittering may be heard when all is well with the world.

Some friends of mine once surprised a mother skunk and her seven little ones in a brushy park. Sensing danger, the mother started off through the brush as fast as she could go, with the seven half-grown skunks close behind her and running almost in unison, nose-tip to tail-tip. She chittered to her young with a sound so soft as to be just barely heard and turned her head quickly from side to side, watching them to see that each kept its correct place in line. It was almost as if she were saying, "Children! Children! Stay as close as you can to me. Move fast, but make no sound. There is danger!"

One time, in the hills of central California, I heard the mating song of the striped skunk, though I did not recognize it at the time. It started as an uncanny whistle in the night—something like the cooing of a dove, only higher; then slowly it became deeper and more guttural, as if going way back in the throat; and here it wavered for a while, going up and down the scale, at last ending in a long-drawn and soft purr like a cat's. I think now that this was the male's call. The sound had in it a quality of deep longing that must surely have touched the heart of any female skunk listening to it.

Skunks make very good pets, especially when they have been descented; they behave very much like cats. One pet skunk I knew always gave a little whine when it wanted to be taken up in the arms of its master and fondled. When it wanted to draw attention to its need for food or to be let out of the house, it would scold with a high, chittering note and sometimes stamp its feet.

Angry skunks have been heard to give a single, low, hoarse bark. At mating time in the hills, much squealing is heard, meaning often that a female is resisting the too ardent advances of a male or that two males are squaring off to fight.

The little spotted skunks of the West are more humorous and friendly characters than their big striped brothers. I have helped to tame wild ones simply by offering food until they came whining to beg it from my hands. They were quick and nervous, and constantly explored everything with wriggling noses.

Spotted skunks are particularly cunning little creatures, about the size of a cat. They are afraid of the larger white-tailed skunks and try to avoid them, which they can do by going into smaller holes and thicker brush than the big skunks can.

One morning I came out to our barn and heard a tremendous, furious chattering from inside a big metal barrel that was almost empty of grain. I knew from the shrill sound that two spotted skunks were in the barrel, and I sensed that they were a male and female and probably mates. A terrific argument was going on in that barrel, and it sounded like one skunk, probably the female, was much angrier than the other; in fact, she was practically yelling at the top of her voice. When I looked into the barrel, I saw that the male skunk was

In a scene of domestic strife, spotted skunks try to escape from a barrel.

crouched in one corner, obviously trying to placate the angry female but not succeeding very well. It was very plain what she was saying; it went something like, "You dumb fool, look at the mess you've gotten us into! Now the man can easily kill us because we can't get out of this barrel!" The male tried to reply, "You know you wanted to get into this food just as much as I did!" Unfortunately, the poor guy could hardly get a word in edgewise!

I ended the argument by very carefully sliding a long, rough two-by-six-inch board into the barrel so they could climb out to safety. I stood by very quietly, and they made no attempt to spray me. But they ran very quickly into a hole under the barn, and I could hear the murmur of their voices for some time.

THE CAT FAMILY

In North America north of Mexico live three common native members of the cat family: the mountain lion, the Canada lynx, and the wildcat, or bobcat. Along the southern borders of the United States, some rarer, wild cat creatures are found, including the big spotted jaguar, the lesser spotted and streaked ocelot, and the small brown or gray cat, with slender neck and head, called the jaguarondi. But my story here will be about the three common cat creatures.

There are certain basic instincts and ways of communication that are common to all our native cats. All of them hunt by the stalking or waiting methods, rather than by the chasing method of dogs and wolves. Both stalking and waiting methods of hunting end by a pounce or a series of swift bounds upon the prey. When stalking, the body of the cat seems to concentrate into the smallest possible space. Every bit of shelter is taken advantage of, and the direction of the animal being stalked can be told by the point of the cat's nose and eyes. The lashing or bobbing of its tail tells us that it is warning all other hunters to leave this prey alone!

When the cat freezes into silent immobility, we know that it is afraid the animal being stalked may see the hunter. It moves forward only when it is sure the victim is looking the other way. We realize it is about to make the final jump when it stops, crouches low, and begins to bring the powerful muscles of hip and shoulder into coiled bundles of energy, ready to hurl their owner forward like a suddenly released spring. The tail moves quickly back and forth, the body sways or trembles a little, and then, as a bloodcurdling growl or scream rises from its throat, the big cat shoots forward like an arrow from its bow!

The waiting method of hunting is somewhat different. The cat finds a branch of a tree or ledge of rock that overhangs a game trail, or, in the case of a bobcat hunting mice, simply crouches in the grass near the mousehole. The emphasis now is on patience. It waits in complete silence, sometimes for hours at a time.

If you should be fortunate enough to discover a cat hunting by this method, you will probably notice it only because of the jerking motion of the tail. I once saw the tip of a jaguar's tail moving in such a fashion, in the dense jungles of Panama, but it took over ten minutes of patient watching before I could detect the body of the big cat, so utterly quiet was it and so well did its golden tawny and dark spotted hide merge with the light and shadow of the forest.

You can tell when such a cat thinks that its wait is nearly over and that its prey will soon come. The hindquarters lift ever so slowly, and the muscles begin to bunch along the thighs. The ears are directed sharply toward the sound of an animal approaching, and the body begins to sway very slightly. This swaying motion apparently helps trigger the muscles for a quick getaway. Suddenly a growl or scream sounds out, and in the same instant the big cat seems to disappear, so quickly does it leap forward to the kill.

Wildcats

The wildcat of North America is about the size of three or four house cats and has a voice at least three times louder than a domesticated cat's. The night I heard and watched two wildcats fighting in the dark woods of the Napa Mountains in California, I felt as if my ears were going to split. The yowls, screams, growls, moans, and caterwaulings, as the tide of battle swung first one way and then the other, would have lifted the hair of an Apache.

These loud voices that vividly tell us of the emotional state of the cat, including pain, rage, and fear, are useful to the animal in bluffing its enemy and also in paralyzing the will to escape of a smaller animal when it is pounced on. But the wildcat can also be extremely silent when it wants to be. I once watched one in the hills near Orinda, California, trying to find a bird's nest to rob. It crept through the brush more silently than a ghost, keeping to the dark side of every bush, and crouching so low to the ground that it seemed to worm forward rather than walk. At the slightest noise, it stood perfectly still for moments at a time. Its whole attitude showed me it was trying to avoid discovery at all costs. I wondered why—until I saw what happened.

The wildcat had found a thrasher's nest in a high bush and was reaching up through the branches to grab the tiny birds inside when the parents discovered the cat. Instantly they let out cries of rage and darted fearlessly at the intruder. At their cry, other birds—jays, titmice, wren-tits, towhees, and others—came rushing to the rescue. The wildcat was surrounded and mobbed by screaming birds on every side. With tail pressed close to its body (a sign of depressed spirits) and ears pressed down on its head (a sign of irritation and anger), it slunk away, still silently, through the brush, its black ear tufts twitching. So defeated did it seem that a pair of jays were fooled into diving vaingloriously to within a foot of its nose, screaming names at it. This was a mistake—for the wildcat, a magnificent male in glorious spring colors of tawny orange and black, suddenly lashed its short tail from side to side (a sign of imminent action) and leaped straight up in the air with claws spread like fishhooks. One of the unfortunate jays was brought down in a cloud of blue feathers and quickly killed.

Wildcats mate in the northern sections of our country in late February or early March. Then the woods may be filled with their "singing" and challenges, though these sounds are little heard in areas where they have been severely hunted. Two males of equal size approach each other with utmost caution—their lips writhing up to turn their faces into masks of fury, their squalls, moans, and growls threatening immediate death. Sometimes one is bluffed into suddenly leaping away, its tail down in the signal of defeat. At other times they wail, caterwaul, and growl for many minutes without action and then leave with mutual respect for each other's bluff. But often they engage in desperate battle.

A wildcat sometimes leaves its droppings on top of an anthill, and this may be a signal, to other wildcats, concerning its sex and attitude. Another sign to other wildcats is a scratch on the bark of a small tree as far up as the cat can reach. This mark indicates the size of the animal that made it and may help in scaring away other wildcats from its territory or in attracting a female.

Young wildcats are born in the late spring and act in all ways like the kittens of domesticated cats. Though very appealing when young, they soon turn into the almost untamable wildcat that spits and growls at all who come near it. The mother gives a sharp snarl to warn them when danger approaches and sends them scurrying into their den in the rocks or a hollow tree.

Lynx

The large cousin of the wildcat in the far north, the lynx, is very similar to the wildcat in habits and attitudes, though it is an animal of the deep pine woods (the wildcat's home is in the brush, broadleafed woodlands, or abandoned farmlands). It is much better adapted for running about on the winter snow because of its large, flat feet.

At the time of mating, in the latter half of October, there is considerable caterwauling, yelling, and growling by the males, especially on moonlit nights; it sounds like many battles are in progress, but actually there is far more noise, yelling, and bluffing than actual fighting. Proof of this are the numerous unharmed lynx skins brought in by trappers. Probably the larger animal simply bluffs the smaller one away by his loud noises and by puffing out his body.

The spring litters contain one to six young lynx. The kittens are noted for their fierce and penetrating meow when hungry. By midsummer the young are weaned by their mother, and by fall they are large enough to move about on their own, though bands of four, five, and six are often seen wandering and hunting together at that time. When such a lynx band is hunting the snowshoe rabbit, members may spread and actually drive the rabbits as men do, calling signals to one another with sharp whistling noises. They also watch each other and, when one lynx is seen in pursuit of a rabbit, the others circle about to spring upon it as it runs past in its fright.

It is possible that the tall, black ear tufts of the lynx, as well as its nervously twitching black-tipped tail, help in signaling. The ears are laid back when the lynx is about to spring, and at the same time the tail waves furiously. The ears, when cocked forward, tell another lynx to wait and watch and not disturb things until the hunted animal is located.

Lynx watchers report a sort of yowling mating song, a weird cry that is halfway between the long "Hal-loo-oo!" of a woodsman and the wailing cry of a loon. It starts with a long, deep "Me-ow-oo-oo!" followed by several cries of a shorter "Me-ow!" It rises in pitch and volume into a continuous "Row-row-row!" wail, which gradually turns into a series of hair-lifting, nerve-racking screeches.

Mountain Lions

Silent, almost invisible, utterly wary of man, the mountain lion, or cougar, sneaks among the rocks and brush of a hillside, merging its long, brown body with every bit of cover. It is looking for signs of deer, but long and sad and terrible experience has taught all of its kind that they also must constantly watch for the guns, bullets, and dogs of men.

In my many years of wilderness traveling, I have seen mountain lions in the wild only four times. Once was in the jungles of Panama.

A mountain lion prepares to pounce on its prey.

The Meat Eaters

In the Trinity Mountains I caught a weird glimpse of the great head of a lion who was following me, outlined against the full summer moon. The curiosity, rather than hostility, of the lion I sighted was shown by the way the head and ears were cocked toward me, in an attitude that plainly said, "What in the world are you doing alone up in the mountains at this time of night?" Had the ears been laid back and the tail lashing from side to side, I would have been in grave danger. Lions very rarely follow men to attack them; in almost all cases it is because they are curious.

Because it is so often hunted, the mountain lion has learned to use its voice as little as possible, though its noises, when it makes them, can be quite awe-inspiring. I once heard a mountain lion scream in the wild, and have twice heard the weird, womanlike scream in zoos. Heard on a dark night in the mountains or woods, the cry of the mountain lion could make your hair stand on end. Even in the zoo, the shrill, trilling, and startling sound, commencing low in pitch and gradually rising to a crescendo, caused my heart to jump with alarm. The cry seemed part human, part angry, and part questioning—altogether uncanny. It expressed an emotion that was half the longing of

A female mountain lion with her cubs.

Mated mountain lions.

a male calling a female, and half the angry determination to keep other males away.

I have seen two mountain lions growl and spit at each other. Sometimes, in the wild, the males get into terrible fights over a female, though often by his ferocious noises and snarling face, the bigger animal bluffs the smaller into running away. The female at first snarls, snaps, and swats at the male who approaches her, but, if she is in mating condition, she gradually allows him to come closer until the two are licking each others' bodies, rubbing together, and purring like two giant tabbies. A male lion may call to his mate with a kind of "Whoo-ee!"—probably asking her to join him. He uses a terrible cry, between a scream and a growl, to warn her of danger.

The mountain lion usually stalks or lies in wait for deer, which are its principal prey. The lion I saw stalking a buck in the forest of the Yolla Bolly Mountains in northern California inched forward with great care, only the black tip of its tail moving—possibly to signal other lions that this game was spoken for. The ears lay back just before the spring, and all the muscles gathered together for the great leap. As he leaped, he screamed to make the buck freeze, and his claws raked the buck's rear; but it escaped.

WILD DOGS

The common native wild dogs of North America are the wolf, the coyote, the red fox, the gray fox, and the kit fox. Some of the actions and languages of these animals are very similar to that of the tame dog domesticated, but most are more typical of wild creatures. In traveling, for example, a dog is apt to go in a straight line, while a coyote or fox, unless in a hurry, will investigate every bush or clump of grass for a possible mouse, rabbit, or squirrel. Grown wild dogs rarely show the wriggling, puppylike attitude of domesticated dogs, though they do wag tails as a sign of friendliness.

Since members of the dog family depend largely on smell for tracking down their prey, it is natural that smell has a far richer meaning to them than it does to cats, who depend more on sight. The nose of a wolf, coyote, or fox is a delicate instrument that sifts out the tiniest smells from the breeze and reads from them many messages of vital interest.

Most members of the dog family leave signals in the form of smell stations at strategic places; they simply wet a bush, a tree trunk, a clump of grass, or a rock. After wetting, the animal, if a male, usually scratches hard on the ground with the hind feet, which may also leave a message as to his size. If a member of the same species comes later to this smell station, he sniffs it all over, gets the message, and then wets it down to leave his own message.

You can often see what message he is getting if you watch closely. If a male shows great eagerness over what he smells, and if he whines, usually he is smelling a female who is lonely. If the animal shows indifference, the message probably tells him that the other is a much smaller male or a female that is not potentially interested in him. If he bristles, flattens his ears, growls, and scratches the ground strongly with stiff hind legs, he has probably caught the smell of a dangerous male.

For animals that hunt in packs, such as the wolves and occasionally the coyotes, the smell stations may act as gathering signs for the pack, the animals following the smell stations until they find the pack. Newcomers have been known to roll their bodies in this scent in order to make themselves more acceptable to the pack.

Wolves

Wolves are probably the most highly organized mammals of North America, after man. A wolf pack on the trail of prey is a thing of machinelike deadliness. After the prey is chased for miles, until it is too tired and must turn at last at bay, the pack closes in on all sides with the scientific timing of a championship basketball team. It hamstrings the hunted animal before it knows what has struck it, tears open the throat, and brings it to the ground, all in a few swift, slashing movements.

The wolf pack is a good deal more efficient than even the best pack of hunting hounds, largely because the wolf leader is right there with the pack, while the true leader of the pack hounds is the man who follows far behind them. The wolf leader gives commands almost entirely by the action and appearance of his body and face, and any disobedience or stupidity is swiftly punished by a slash of terrible teeth. When the leader feints as if to grab for the leg of a buck deer, this is often a signal for another wolf to make exactly the same feint from the other side. As the deer whirls frantically to protect both flanks, a third wolf sees a hind leg coming near him, and dives in and slashes through muscle and sinew with one powerful snap and jerk of his mighty jaws, hamstringing the deer.

The male and female wolf, who generally mate for life, usually very intelligently and intensively rear their cubs to prepare them for life in the pack. Instant obedience must be taught at an early age, as this is the law of the pack. A naturalist once watched a father wolf leading his family across the tundra of northern Canada. The mother wolf acted as rear guard, while the cubs were strung out in single file between. The father, who was about two hundred feet in front of the cubs, reached a rise of ground and stopped to look ahead. Back and forth swung the long wolf nose, testing the air for smell. Then he swung his head around and looked at the cubs. Not a sound was made, but the three young ones stopped still in their tracks as if they had been speared to the ground. The big father watched and sniffed

A subservient wolf appeals for mercy in a fight.

for a while, then swung his head back to look at his children once more. Something—the way he moved his head, or the curve of his lips, or the look of his eyes—told them everything was all right, and they trotted on once more.

The mating season of wolves is generally between January and March. It is during this time that the females choose their mates and the males battle for mastery. But the female does not always choose the male who has won the fight: she may join the loser and drive off the winner.

The preliminaries for battle are similar to those of dogs. But in the wolf's fight there is even more effort to oppose teeth to teeth. As with dogs, if one wolf feels he is losing, he can beg for mercy by suddenly exposing his throat to the other.

Beyond the language of movement, wolves have a more complex language of sounds than do most other animals. European wolves have been reported to have ten sound groups, each conveying a wide variety of meanings. Let us see if we can describe the meanings of the chief sounds of American wolves.

1. Young wolves whine for their mother when they are lonely, and she may answer them with an inquiring whine. It is as if they cry, "Mother, Mother! Where are you? We are hungry!" and she answers back, "I am coming with food. Are you all safe?"

2. The bark of the young is either a part of the language of play—hide-and-seek and so forth—or, when given on a higher note and accompanied by the whine, a desperate cry of hunger. If it becomes too loud, the returning mother hurls a short, deep growl at them, meaning "Silence!" There is generally instant obedience. Wolves also may bark as a command to follow, meaning "I have found something! Come see it."

3. The growls of wolves, especially the big males, are usually coarse and deep. Many dogs are so intimidated by the fierce threat of this growl that they dare not attack a wolf. The growl is also used by the wolf, along with the snarling face and raised hackles of hair, to bluff a lesser wolf.

These growls may be used by the mother and father wolf to warn the cubs to lie still or to come forward, or for other commands. Usually no second growl is needed, as the cub knows he will get a severe nip if he does not obey. The growl of the mother is also used to warn strangers away from her cubs and den.

4. A snarling whine is used by the cubs in play and during mock battles. It is high-pitched and rises still higher when the cub is either hurt or angry. It may mean the cub is in high spirits and is trying to dominate his companions.

5. A musical, long, smooth howl (similar to that of large dogs of wolf ancestry) is the mustering or rallying cry and is a wonderfully stirring thing to hear on a winter night in the northland. Most often the leader of a pack stands at a high place in the forest and throws forth this far-carrying cry.

6. As the gathered pack sweeps together through the dark forest, a series of high-pitched howls rise from them, vibrating on two notes. This is the hunting song of the pack in full cry on the trail of a deer or caribou. To the frightened game, it is a song of death.

7. Suddenly the pack breaks into a louder, different note, for they have caught sight of their prey, running ahead of them over the snow. There is a deep yelp of eagerness in this howl, almost a yammering, and an expression of bloodthirsty desire. The poor deer is frightened

into redoubling its energies and plunges forward recklessly. This is exactly what the pack wanted when it gave its yell, for the new surge of energy soon tires the deer, and thus makes it an easier prey.

8. Now the frightened deer begins to stagger and the pack swings out to surround it, giving a series of short, sharp barks and howls that is the killing song. First, this song tells the leader where each of his followers is so he can more easily direct the attack. Second, it frightens and confuses the deer more than ever and so makes killing easier. In a few seconds the pack closes in, avoiding the flying hoofs. Feint to the right, feint to the left, then strike—and the deer, hamstrung on both hind legs, sinks helplessly to the ground. It is usually the leader's privilege to jump for the throat and have the first taste of blood.

9. Sometimes a howl is heard that seems to sound like all the loneliness of the world. It comes from a lone wolf on a far hill. In it is something of the rallying cry, and the hunting and the killing song, and something else. It is a song of despair, usually sung when the wolf feels all its companions are dead and it alone is left in the world.

If a wolf wishes to join a strange pack, it approaches with waving tail. It may even act playful, like a pup, and turn over on its back with legs waving in the air when the pack approaches. Sometimes this works, and it is accepted into the pack after a few perfunctory bites to try its mettle. At other times it is harried unmercifully, bitten and rolled over and over by bullies in the pack. And on some occasions the pack has quickly killed the strange wolf.

Coyotes

The coyote is a more appealing animal in many ways than the wolf, not so untamable and far more adaptable to the presence of man. Whereas the wolf has retreated before man and his guns, traps, and poison into the farthest wildernesses, the coyote has spread to new places because of the opening up of the country by the felling of timber, and possibly also because of the destruction of the wolf, who is a deadly enemy of the coyote.

Coyotes only occasionally gather to hunt in small, loosely formed packs that are not as efficient or as large as those of the wolves. Their habits are even more doglike than the wolf's, but like the wolf they usually mate for life, and the father helps take care of the young ones, something rare among domesticated dogs.

One coyote father that I watched in the San Francisco zoo was extremely proud of his children and stood guard over them with great vigilance and courage. When a visitor approached the bars, he would rush forward with bristling mane, snapping teeth, and a savage growl deep in his throat. If he thought there was real danger, a sharp bark of warning would send the young ones scurrying for the rear of the cage, where they would cluster behind the equally bristling mother. The father looked at the young coyotes as if they were the most wonderful things on earth. When he was sure conditions were safe, he would lick them around the ears and neck with a proud paternal expression, as if to say, "These beautiful creatures are really mine!"

In the wild, when I have been in the neighborhood of a coyote family, I have heard the mother give a long-drawn-out and very quavering squall as a warning to her young. Sometimes two or even three mothers have dens together, or a grandmother may come to help with the young ones.

Out of such groupings of families and possibly of other closely related individuals arise the small coyote packs that sometimes gather together to hunt jack rabbits or even attack a wounded or sick deer or a calf they have separated from a cow.

These packs like to get together on beautiful western nights and sing to the stars or the moon. Their singing, in simplest form, is a series of yaps, almost like laughter, followed by a long squall. The sound is often ventriloquial: you seem to hear them calling from one hilltop, when actually they are singing on another one a mile away. When several coyotes join together, the sound becomes quite tremendous in volume. It is filled with a wild longing that is enjoyed by every true lover of the outdoors. Often an amazing combination of whines, barks, howls, and wails rise and fall in the desert night. One night, as I listened in the Arizona desert, it seemed as if the sound came from every side of me, sweeping down from the dark buttes and hilltops like a chorus of surrounding demons, and trailing away finally in a last, long, fading howl.

The song of the coyotes when gathered together is very much like a "social sing" of men, and expresses high good spirits and the joy of companionship. But a single coyote male singing from some high hill

A coyote howls to his pack or mate.

is likely to be trying to find a mate. Then the sound seems lonely and wild, filled with deep, sad notes and a sense of laughter through tears. The females are no doubt attracted to this song, if well done—even as human females are attracted to a romantic male singer.

Coyotes are famous for their wisdom, cleverness, and cooperation in catching animal prey. A coyote male touches noses with his mate and by this signal tells her to sit down and wait for him to call her. Then he marches, singing merrily, through a prairie-dog town. The little rodents immediately dive into their holes, and, as soon as they do, the male gives one sharp bark. At this signal, the female rushes forward and hides behind a bush, near a prairie-dog hole. The male takes up his song again and walks on through the town. As soon as the prairie dogs hear him singing off in the distance, they think all danger is gone and pop out of their holes to see what that crazy fellow was singing about. Instantly the female coyote pounces on the dog nearest her, and so the wise ones get a meal!

Two or three or even more coyotes often join together to chase

Coyotes lie in wait for sheep.

down a jack rabbit. There is little need for more than the language of motion, as they all know the jack will run in a great circle. After a jack rabbit is started, one coyote chases him, while the other two watch carefully to see which way he will circle. As soon as they are sure, they trot off across country in such a way as to intersect the circle. When the running coyote begins to get tired, he gives a series of sharp barks, and the next coyote sneaks through the brush toward the sound, so that when the frightened jack comes near, he can spring out on him. The jack dodges the slashing teeth like a flash and runs on, but he constantly has a fresh coyote on his heels, while the other attacker rests by walking slowly toward another point in the circle. Round and round the jack goes, until, utterly exhausted, he is an easy prey for the last coyote who chases him.

Coyotes, like wolves and dogs, leave their scent signals on bushes and in mountain paths. From these a newly arrived coyote reads the stories of the others who have gone by before him.

Red Foxes

The elegant and clever red fox is one of the most charming and interesting animal characters of North America. In the woods of New Jersey, a friend of mine once watched one stalking mice. Unaware of my friend's presence, the fox moved through the grass with the grace of a ballet dancer, each prettily furred foot put down with such care that not a sound was made. When the mouse was sighted, the sharp nose pointed directly to it, the ears cocked forward eagerly, and the fox began to ooze forward so slowly that it hardly seemed to be moving. Suddenly the muscles of the thighs bunched tightly, the ears drew back, and the red body hurtled forward. There was a sharp snarl, a squeak—and the jaws snapped together on something soft.

In captivity the red fox shows a marvelously expressive face that is a key to a very cunning and complex character. In the wild this character is clever in its ability to live successfully in and around the farms of men. The abandoned farm, where cut-over and once-plowed land is growing up again with bushes and grassy meadows, is the red fox's favorite hangout.

The face of the red fox undoubtedly signals to its own kind how it

Red foxes.

feels. In a snarl it expresses enough fury to bluff a rival. The black lips draw back from the red gums and sharp white teeth. The eyes become slits of rage.

The fox face has also been observed laughing at the bungling of a dog or man. In such a case the lips seem to stretch in an actual grin, and the tongue hangs out and shakes. When expressing alert watchfulness, the ears cock forward, then sideways, and the whole face takes on an attitude of animated curiosity.

The voice is almost as varied. The cry of the male is notably coarse and heavy, while the female's cry is usually very shrill. A short bark, followed by a little squall-like "Yap-yur!" is apparently the hunting cry of early evening, a warning to other foxes to stay away from the territory being hunted.

A long yowl is probably the male's call to a female at mating time, as it often has a note of loneliness in it, and the female answers with a characteristic shrill squall. As the two sounds grow closer together, they may change gradually into sharp eager barks from the male and answering squalls or barks from the female. The female may repel the male at first with snarls and bites before she decides she likes him.

Red foxes have been known to give two or three different screeches. The meanings of these are clear once you have observed the foxes' emotional moods. By the tone of the sound, it expresses anger, fear, pain, or warning. Red foxes have been known to screech and squall to rouse dogs to chase them—a form of teasing. The fox may occasionally yap to encourage the dog to keep on following.

On moonlight nights foxes have been known to chase each other about, barking loudly, and making soft churring noises of good fellowship. These sounds are far different from the coarse and savage screeches and snarls of two male foxes approaching each other to give battle.

One of the most sinister and unearthly noises of the North American wilderness is the terrible scream of a male fox when he is being attacked by superior forces or is trapped. This frightful noise may bring help or may bluff a large enemy. The vixen (female fox) has been known to squawk and squall like a heron in similar circumstances.

The mother fox gives a sharp bark when danger nears, to warn her young to dive into the den and to be quiet. She may then circle away with many yaps, trying to draw the enemy into following her and leaving the youngsters alone. Later, if successful, she will return in triumph to her home and call the young out with a soft bark or churr.

Though the red fox has extremely sharp ears and eyes for detecting signs and smells even at far distances, it apparently cannot see a man if the man is standing still. If you see a fox and a wind is blowing, carefully try to get behind the wind so that it comes to you from the fox. Then stand perfectly still. Watch and listen for either a meeting of foxes or the sight of a fox stalking its prey. The stalking is done with great cleverness and skill by an older fox, but more clumsily by a younger one. A fox signals other foxes by its high-pitched bark and by smell, from a scent gland in the tail, which is placed along the trail and tells other foxes what kind of fox has passed that way.

Fox cubs learn from the older foxes, but the young fox is quite clumsy at first and makes many mistakes, whereas such simpler creatures as mice and rabbits instinctively have knowledge for escaping enemies almost from birth. As a fox cub grows older and is taught by its parents, through voice signals and nips, how to avoid dangers, it becomes more intelligent in its choices and finally can be trusted to find its own way in the world. The father and mother constantly train the young by sharp nips when it does wrong, by sharp barks of command, and by smiles on their faces when the cub has done well, along with a nudge now and then to get them moving or prompt them to do something better. How to actually stalk a mouse or rat is shown over and over, especially how to grab it so as not to get a return bite. The hardest thing to teach a cub may be the ability to sit absolutely still and watch.

Foxes vary in intelligence just as humans and dogs do. Some very clever father and mother foxes may show the young ones how to smell out and uncover a trap, and even how to cause the trap to snap. It is vital for foxes to have a special signal to tell the cubs that *they* are coming to the den and not some other animal. The mother or father give a long, soft, murmuring "Mmmmmmmmmm." This is another example of fox intelligence. When the cubs hear the parent, they come tumbling out of the den, eager for food and play.

One very remarkable way a red fox has for catching food is to toll, or lure, ducks to the shore of a lake by acting crazy. As the ducks approach the shore to try to find out what it was being crazy about, the fox hides in a nearby bush. When the unsuspecting victims get close enough, the fox rushes out and catches one of them. Clever fox!

Foxes very often hunt mice. If you are downwind in a meadow where a fox is hunting, you can kiss the back of your wetted hand or use a wet leaf to make a squeaking noise that will lure a fox, if you sit absolutely still. The mouse squeak is a signal to the fox that the mouse is not aware of danger but is "talking" to another mouse.

The cleverness of a fox is shown by the way it outwits other animals. For example, naturalists in a small plane saw a fox come upon a deer that a hunter had killed. A badger was scavenging the deer and wanted it all to itself. As a badger is an extremely touchy and dangerous animal, the fox decided to draw the badger away from the deer. He harassed the badger by dashing at it and pretending that he was going to eat some of the deer. The badger would chase the fox, and the fox would lure it far enough away from the deer that he could rush back, being much faster than a badger, and grab some meat. The fox did this over and over. This communication between fox and badger was to the badger's detriment.

The scent posts of foxes are made to send stories between two or more foxes. The scent of the male is usually put on a bush or bit of high ground, while the scent of a female is placed in flatter places. The scent of the female tells the male fox whether she is ready to mate. The scent of the male at a scent post tells the female he is looking for her. Two males vying for one female may get into a severe fight, accompanied by considerable snarling. The bigger one usually succeeds in chasing the other off, but the female may not accept him.

The tails of foxes are used for communication with other foxes. A male who holds his large tail high and spreads the hairs out to make it look bigger may actually outbluff a fox with a smaller tail and make him run away without a fight.

The voices of foxes are usually sharp yaps, telling other foxes of their whereabouts or calling young foxes back to the family, but occasionally they give bloodcurdling shrieks. What these are for we do yet not know. The female is the one that usually gives this kind of shriek.

Gray Foxes

I have watched the gray fox in the western wilderness several times, delighted with its quick motions, expressive face, and appealing character, but this kind has little of the cleverness of its red cousin. The gray fox runs straight away from a chasing dog, with few of the tricks, like running in water or doubling tracks, so often used by the red. The gray usually heads straight for impenetrable brush or rock piles that he thinks a dog cannot enter. Sometimes he crawls into an abandoned badger hole or climbs a tree with a sufficiently slanting trunk. He has even been found sixty feet up in the nest of a hawk or crow!

In the winter, the short, sharp, hoarse barks of the males in mating mood are answered by the shriller barks of the females. The voice of the gray fox is coarser and deeper than that of the red fox, and the males are noted for vicious fighting. Messages, as with the other foxes, are left at scent posts along the trails. The cries of the young and the warning cries of the adults are similar to those of the red fox.

The gray fox sometimes gives a high-pitched, soft whine, almost inaudible, when it catches sight of an animal it wants to eat, and this may be used as a signal between a hunting couple.

Another cry is a miserable-sounding harsh squall, used when the fox has missed a mouse or rabbit for which it jumped, or is otherwise frustrated or angry. I kept two little gray foxes in captivity once for several weeks and heard this squall once when they were very hungry. Their faces were always very expressive—whether of mischief, anger, curiosity, or some other emotion—and they were extremely fast and nervous in their reactions. I was glad to let them go free.

Kit Foxes

Under the full moon in the desert in Death Valley, California, I have watched kit foxes come down to the edge of our campfire light and skitter back and forth like little gray goblins, waiting for a bit of food to be thrown to them. The soft barking of the kit fox in the desert night is a magical sound. This barking becomes louder in the mating season, around the first part of February; and with it may be heard a cry like an owl's screech. The barking seems to be a signal that food has been found or a territory call; when given with an angry tone, it is a challenge from one male to another. The screech is probably a cry of angry bluff when two males meet in rivalry over a female. Another screech seems to be given when the fox misses an animal in a pounce or when some other hunter seizes its food. The ears flattening down are a sign of danger, as is the sudden dropping of the tail.

SEALS AND SEA LIONS

The common earless seal of our coasts, also called the leopard or harbor seal, grows to be about six feet long. These seals mate indiscriminately, without family life except for the female with her young, but they do like to gather socially in large and small herds that travel the seas together. The association seems to be rather loose, and there is little cooperation observed in hunting for fish, though no doubt the excitement of one leopard seal as it finds a school of fish is quickly communicated by its actions, the sight of which brings other seals to join in the chase.

As these seals lie on the rocks or ice, certain individuals appear to take on the job of sentinel. When danger approaches, a chorus of loud, short, and sharp barks sends all the nearby seals bouncing into the water.

The bark may be a challenge and a warning to others to stay away from a particular territory. It is also used for social communication, as the seals travel together through the sea. Another warning as well as challenging sound is the deep snort of the male, which is also used to impress nearby females at breeding time.

Baby seals give a cry that sounds like "Mama!" when they are hungry or lost—a baaing almost like that of lambs. They give a more desperate "Kroo-oo-ah!" cry if Mama does not come when expected. Each mother knows her baby by scent, which she absorbs when she licks it all over at birth.

The sea lions, which I have watched for hours on the rocky islands and beaches of the California coast, are far more fascinating animals. The great male, two or three times as big as the slim female, has a huge, muscular back and shoulders like those of a grizzly bear, and is lord of a small harem of one to twenty females. In the spring breeding season, the bellowing of the males as they fight can be heard for miles. Each male stakes out his territory on the rocks and attempts to corral a group of females.

The bellow of a male sea lion turns into a loud roar when he rushes at an enemy to drive him from this territory. The two move their heads like striking rattlesnakes, each trying to seize the other by a flipper or an ear. Often the battle goes on for many minutes with only

A California sea lion colony on a beach.

the clash of teeth against teeth until one animal, badly wounded on a flipper or around the head, gives a bellow of despair and turns and runs for the sea. The winning bull sends forth a deep, hoarse bellow of triumph and proudly tucks his head against his shoulders while the nearby females look on admiringly, possibly barking softly among themselves.

The female has a weaker, shriller cry than the male, but it is still quite loud, and a whole herd of sea lions talking to one another makes quite a din! By watching the animals carefully as they roar and bark and whine, you can gradually begin to make something out of the confusion. A big bull roars a protest because a young seal has bumped into his tail while he was snoozing. He may lift his whole body up and sway in the air while he roars, or he may turn swiftly around and dart his head, with its big dog teeth, at the disturber.

A shriller bark from a female calls for her baby to come out of the sea and get up on the rocks. You can detect the worry in her voice. She thinks there may be killer whales or sharks near, and she doesn't think he is old enough yet to take care of himself. With a whining bark or mew, a youngster calls for its mother's milk. And over on one rock, two young bachelor bulls are having an argument, during which they practice bellowing with all the viciousness and ardor of big males. Sometimes a bull, or even a cow, may snort a warning that will cause every sea lion in the sea to come rushing for the rocks. The message of the sound is clear; it means "Killer whales!" Ever since I saw the sea turn red with sea lion blood near the Farallon Islands, I have understood the sea lions' almost hysterical fear of these great, black-finned wolves of the sea.

While the eared seals, such as the sea lion and fur seal, can move about quite agilely on the beaches and rocks because of their jointed hind flippers that act like legs, it is in the sea that their true beauty of movement is observed. Down into the darkening surface zones they go like diving bullets, chasing the fish of the depths, a row of bubbles streaming from each nostril. Below the sea surface, the language is one of pure movement, the direction of the driving, churning body and flippers telling other sea lions not only that fish are here, but also that the particular fish this sea lion is chasing belongs to him alone.

Captive female California sea lions are the ones most often seen performing in zoos and circuses, because of their considerable intelligence and the ease with which they learn tricks. They bark beggingly for food, clap their flippers proudly when some special trick is accomplished, and make enormous leaps from the water over obstructions—only if a good reward of fish is furnished them at the end.

THE BEAR FAMILY

Bears are built for strength rather than speed and so are eaters of about everything they can find, including vegetable as well as animal food. This omnivorous diet makes their temperament and ways somewhat different from those of the single-minded meat hunters, such as the weasels, cats, and wild dogs. They are inclined to be more lazy and placid and greater seekers of comfort—especially during the winter, when they go into hibernation, while the other animals must struggle to stay alive.

Bears are extremely shortsighted creatures, which means they live mainly in a world of scent and sound. A bear, when it sights a man, often starts to circle upwind to catch the scent and discover just what this creature is. Then—with a loud "Whoof!"—it wheels and runs.

Black Bears

The black bear is a somewhat more jolly and comical fellow than the usually dour grizzly. It is not above acting like a clown, especially if this may win some extra food. Like a grizzly, it claws the bark of trees as high as it can reach, probably to show strength and size, as well as to sharpen claws. You can tell a black bear's mark by the sign of four claws. A grizzly gashes the tree with five.

Besides leaving claw marks, the black bear (the brown and cinnamon are simply varieties of the black) often plasters a tree with mud and rubs against this mud with its back, leaving both hair and bear smell. This is a complete bear-identification message for the next bear who comes along. Male bears, in particular, study these "bear trees" carefully, and then try to claw higher than those who were there before them. The highest mark may warn others that a really big bear is in the neighborhood and that smaller males had better clear out.

Every wild bear I have met in the mountains of northern California gave that startled "Whoof!" as soon as it smelled me and rushed pell-mell into the brush. The noise, as heard by other bears, would certainly mean "Danger!" One cub I met quickly climbed a pine tree and disappeared among the branches. It was very quiet, which led me to believe its mother was not near.

Many people are injured every year in the national parks because they become too friendly with the park bears and do not understand their language. When such bears lose their fear of men, they become first-class highway robbers and night marauders. In Yellowstone Park, I spent most of three nights chasing bears away from our canvas-covered trailer. Once a mother and two cubs climbed on top of the trailer and started tearing at the canvas. Only by flashing a strong flashlight in their eyes and yelling at the top of my lungs was I able to drive them away. Growling, with ears laid back, they edged off, since my bluff was louder than theirs!

Tourists who stop to feed bears on the highway in the national parks are courting danger; they do not realize that the bear is a large and savage animal who, when its appetite is aroused, is motivated to take everything it can get. Such a bear may try not only for a sandwich that is being held out to him, but also for the hand that is holding it! Holding back food from a hungry bear is even more dangerous. The tourist will see it lay back its ears—a sure danger sign—and lift its lips in a snarl that shows its teeth. A wise traveler steps on the gas at this point and gets out of there in a hurry!

Like most animals, bears have a language of movement as well as of sound. Most of us have seen the clowning bear at the zoo who sits on its tail end, grabs its hind feet with its front paws, and sways back and forth like an animated rocking chair, craning its head and neck in a pleading motion. This is sheer, shameless beggary, but this bear does profit by it. Soon the peanuts begin to fall.

A playful bear likes to start a game with a comrade by running up and playfully nipping its rear end. The other bear, often too lazy to move much, falls over on his back and strikes at his tormenter with both forepaws. In an instant the two are rolling over and over in an ungainly wrestling match that is mainly pure fun but can turn into a semiserious scuffle if one of them bites or strikes too hard. In watching this play fighting, you will seldom see anything like the so-called bear hug. A bear does not crush its enemy with a hug, but may reach

A female black bear sends her cubs up a tree.

out and grab with the forepaws in order to bring the other animal close enough to bite.

Bear mothers seldom spare the rod and spoil the child. If a bear cub does not get up a tree fast enough when told to by a grunt or growl from Mama, he is liable to be knocked halfway up the tree with a blow! There is good reason why the lesson of discipline must be taught quickly, because there are many animals lurking in the woods that would be happy to gobble up a tender young cub. Soon the cub learns that a hoarse grunt from mother bear means for him to climb a tree—and fast.

The mother is quite affectionate with her cubs when in the mood, and grunts fondly when she finds a lost cub, licking him all over. Cubs whine and whimper loudly when they are hungry, but are hushed up in a hurry by a low growl if danger is near. A peculiar, whimpering, high-pitched moan is given by a cub to call its mother when it is lost.

Once I heard a caged bear make a prolonged murmuring, whimpering sound, and thought at first it was sick or hurt. But it was lying luxuriously in the sunlight, and its later actions showed it was quite healthy. Apparently the sound was one of pure contentment and well-being.

Many other sounds give very different meanings. There is the growl that is a threat and a bluff. There is the loud cough of menace when the bear is getting ready to attack. This sound is also given when two male bears are about to fight. The bear with the deeper cough and larger size may bluff its adversary into turning and running.

That a bear can be a coward, or perhaps just realize it has no chance in a fight, was shown to one naturalist who observed a small black bear in a fight with a mountain lion. The bear fought back only when actually attacked; otherwise it rocked back and forth with its back to a cliff, moaning and screaming. It was trying to tell its enemy that it did not want to fight at all; but this apparently did not work, because the bear was soon killed.

Some black bears bawl like bulls when in rage or pain. This cry may also be used to ask for help, if a mate is near, and is often heard when a bear is caught in a trap. The hoarse, panting, coughing roar of an angry male bear is usually broken by savage snarls and many a violent "Woof!"

Two male black bears prepare to fight.

At mating time, male and female black bears are often extremely affectionate and conversational, rubbing noses and sides, sniffing at each other, grunting, whining, and moaning softly. Most of these sounds simply express affection in various tones. The same is true of the grunts, mumbles, whines, and squeaks often heard between a bear mother and her cubs.

Grizzlies

The grizzly bear is distinguished from the black bear by its larger size, the muscle hump in the shoulders, and much larger, heavier claws.

Besides leaving marks of five claws, grizzlies sometimes bite out big chunks of bark high up on the trunks of trees.

Grizzly psychology has changed considerably since the coming of the white man and his gun. Now grizzlies hide from men and rarely attack unless wounded, except within the national parks, where grizzlies can be very dangerous. Grizzlies occasionally kill a large animal, such as an elk or moose, by rushing the other animal from some hiding place, frightening it with a terror-striking roar, and grabbing it by the neck to throw it to the ground and kill it. After such a kill, the grizzly eats as much as it can, then rakes some fresh dirt over the carcass. This dirt and the smell left with it warn all other creatures, "Leave this alone! I'll be back to eat soon!" Few animals dare to ignore this warning.

Mother grizzlies talk with their cubs, as do black bears. When the mother wants to leave a dangerous neighborhood in a hurry, she gives a short, choppy, coughing sound that brings the cubs close to her side as she runs away through the brush. Once or twice she rises high on her hind legs and turns to look back, sniffing vigorously to see if the danger is still near.

This rising to sniff the wind is a common signal among bears when they are uneasy about something. Other bears see the first bear sniff and quickly catch the same uneasiness. They, too, rise and sniff the air, swinging their bodies and necks from side to side. If the ears lie back and a low growl rises like thunder in the throat, while the breath comes quickly and harshly, it is a sign of an enemy approaching.

Most of the year, except for mothers with cubs, grizzlies are solitary animals, more grumpy and bad-tempered than black bears. During the mating season, in late summer, the big males become more noisy—coughing, growling, grunting, moaning, and roaring. The females answer in kind, but with higher notes. The coughs—"Koff-koff-koff!"—and grunts are usually the calling notes of the bears, trying to contact each other. Fights between males are similar to those of black bears.

Raccoons

The raccoon is a little cousin of the bears. Like them, it walks with its hind heels on the ground and so indeed "walks like a man." But the raccoon is one of the wisest and cleverest of all animals, particularly if fate and natural ability have allowed him to live a few years. He is the sage of the woods. A wise raccoon may know more tricks for getting out of trouble than even the red fox.

The raccoon often has a sociable family life, but some old males become quite solitary. When two family groups get together, you are bound to hear the querulous "Churr-churr-churr!" of the adults squabbling over food, or the soft "Er-er-er-er?" of a young one begging food from its mother.

The front hands of a raccoon, so much like a man's with their opposable thumbs, no doubt help account for some of the raccoon's cleverness, including the opening of chicken house doors. I once watched a raccoon mother pass some chicken intestines to her two young ones, and the passing back and forth was exactly as humans might do it. Each child took its bit of food and sniffed it over carefully, then took it to some nearby running water to wash it before eating. This washing apparently is to help raccoons swallow the food and not for sanitary reasons, as once thought.

The raccoon's face is very expressive. I once watched one wrinkle

A raccoon rejects an unsavory bug.

its nose in distaste when it found a bug with a bad smell; it spit the bug out with every indication of displeasure. Another coon, seeing this facial expression, also made a grimace and threw away the offending bug. The coon also expresses rage, fear, suspicion, and curiosity with its facial muscles and the way it curls its black lips back over sharp, white teeth.

A coon expresses alertness and alarm over possible danger by crouching low, with its head and ears cocked toward a strange sound, the nose sniffing to catch the smell. It may express curiosity by rising on its hind legs and peering toward a strange object, sniffing deeply. A pet coon loves to explore and, with its clever hands, delights in opening drawers and emptying all their contents on the floor! Wild coons have been known to steal bright and shiny objects from cabins or camps, very much as a pack rat does.

Coon families seem to mark out and claim definite hunting territories, like otters. Two coon families, meeting at a fishing pond both claim, may squabble, scold, and fight half the night—with no one getting hurt very much, but everybody losing out on the fishing!

Ordinary conversation is by churring in different tones, one querulous tone coming from a young one eager to get going on an adventure, a deeper note from a mother telling the young ones to mind their manners or be quiet, and another note, with questioning or urging in it, from two coons trying to decide which way to go on a hunt.

In the mating season, male coons fight to the tune of savage growling and snarling, noises they also use when attacked by dogs. They use their front hands as well as their teeth and have been known to drown dogs by pulling their heads under water in a pool or stream and holding them there. Two male coons approaching each other puff their bodies up and spread out their hair to look as large as possible. This puffing and a deep growling, may bluff a weaker raccoon into suddenly drooping its tail and hair and turning to run.

The mating song of the male is sometimes called a whicker. It is a prolonged and very tremulous "Whoo-oo-oo-oo!" Sometimes it is mistaken for the voice of a screech owl, but it is much more squalling and coarse than the sweet, soft noise of the little owl. This song seems to express love, joy, and vigor all rolled into one, and may be answered by a similar but shriller song from the female, quivering through the night. Coons mate usually in midwinter, often for life, with the fathers helping to raise the young ones.

Coons are very careful in training their young, teaching them tricks to avoid dogs, men, and other enemies, and ways to get food. I once watched a raccoon mother showing her youngsters how to catch fish with their hands. She had to scold them with sharp churrs to make them pay attention, and one could imagine her shaking her head when a young coon lunged for a fish with a sloppy technique.

5
Vegetarian Mammals

Most of the animals described in this chapter have two things in common, they feed mainly on plant food, and they are hunted by the animals described in the previous chapter. A few, such as the bats and shrews and some of the mice and rats, eat insects or even other animals. But they do not have the large, sharp teeth of true carnivores.

DEER

The deer, elk, mountain sheep, and similar animals of North America all have in common hoofs and a liking for green things to eat. Many are noted for the horns they carry on their heads, which are used principally by the males in mating combat. They usually travel together in herds or bands, at least at certain times of the year, in order to cooperate for mutual defense, and from this banding comes a strong need for various kinds of communication. Since they are all preyed upon by the large carnivores and by man, they have found that the price of life is constant alertness.

The deer family in North America includes the white-tailed and black-tailed deer, the mule deer, the elk, the moose, and the caribou (with its relative, the imported reindeer, in Alaska). Some English roe deer have been imported into the eastern states and they seem to be growing in numbers there. Only the best-known species will be discussed here.

Vegetarian Mammals

White-tailed Deer

The white-tailed deer is found mainly in the eastern United States and Canada, though it occurs in scattered areas in the West. In early spring, the older bucks go off in groups of two or three, while the does form small bands of a half dozen or so made up of two or three does and their young of the previous year. In May each doe goes off alone to a hiding place to have her baby. The fawn instinctively stays perfectly still and merged with the colors of the ground when the mother leaves it, but it may squeak if it gets hungry.

During the summer, as the young are growing up, the does may drift together again, seeking the help of others to guard against danger. A routine is followed of resting in the cool shade during the heat of the day and searching for food in the early evening and morning. At such times one doe may stand guard for a group while the others feed. At the lifting of her white tail and her soft "Blaat" of warning, the other deer turn in the direction she is running and flee the approaching enemy. The bucks, whose growing horns are soft and velvety and delicate at this time of year, live separately but may have their sentinels, too.

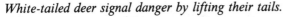

White-tailed deer signal danger by lifting their tails.

Two white-tailed bucks fight.

A deer feeding in a meadow in the forest ducks its head and takes a few bites; then its tail moves and up comes its head while it looks carefully around, its ears spread wide to every sound and its nostrils flaring as it takes in the messages on the forest breeze. At a wrong smell or sound, a deer will snort, raise its tail, strike the ground a blow with its front hoofs, whirl, and leap away. All deer within sight see the flashing of the white tail, and they, too, turn and run in the same direction.

By the middle of September, the buck has cleaned most of the felt covering off his antlers and goes through the forest butting at trees and pawing at the ground, getting ready for the battles of October and November. The fawns are weaned by their mothers this month, and the does begin to run restlessly alone. They are not yet ready for mating.

Bucks at this time of year make soiling pits, or wallows, where they tear up the ground into dust, put their own water in it, and in other ways make a smelly place in which to lie and rub their bodies. These wallows appear to be signal posts and also challenges. Possibly the deer who makes the biggest and smelliest wallow can convince other bucks that he is a dangerous fellow and that they should run when they smell him. Certainly bucks are quite interested in each other's wallows and investigate them carefully. (The reactions of bucks to hunting are described on page 105.)

In October the doe begins to run with her head turned backward,

Vegetarian Mammals

looking to see if she is followed by a buck. If her tail is pressed down close to her body, this is often a good sign that a buck is chasing her and that she is pleased with him. Such deer are still aware of danger, and a commanding "Whoof!" from the buck will send the doe into concealment. A shrill whistle of warning may tell her to run. These signals may also be given by a buck who seeks to keep his doe from being captured by an enemy buck. The buck may also flash his white tail to tell the doe to run and in what direction. She flashes hers as an answer and to show that she is obeying his command. If another buck approaches, the buck challenges with a coarse whistle or snort.

The buck and doe nuzzle each other fondly at the time of mating, and their white tails constantly move up and down as a sign of affection. They may bleat for each other when separated, the buck with a deep, throaty blast that is almost a bellow.

Two bucks who clash over the same doe approach with a ritual in which they stamp and snort and advance a few feet, then give a bawling bleat, then stamp and snort and advance once more. The hair along each back stands straight up. The great shoulder and thigh

Deer: (clockwise from upper left) *a white-tailed buck, with horns in velvet; a buck in autumn; a doe, alert to danger; a doe licking her fawn; a doe sounding the alarm; a curious doe.*

muscles bunch for the charge. But the charge itself, when it comes, is only for a short distance. The horns clash and ring as they meet; then back and forth the two press each other. Each buck watches carefully any move on the other's part that indicates a dive for his flanks. He shifts his horns quickly to meet such an attack. Finally, as one becomes more tired than the other, his tail begins to sag as a signal of his intention; he suddenly disengages his horns and wheels around to escape. In that moment, if he is not fast enough, the other buck will gore him in the side.

A white-tailed deer that passed very close to me one day, chased by hounds, gave a high-pitched shriek of terror as it dodged past. Other calls that have been heard from these deer are a low, muttering sound given by the buck when he is seeking a doe, a scolding cry of the mother to the fawn when the young one wanders near danger, and a soft bleat—almost a mew—from the fawn calling its mother.

Mule Deer

The mule deer of the western United States are of two main kinds. The typical mule deer of the mountains has very large ears and a light brown color. It usually carries its tail dropped and swings the black tuft back and forth over the white patch. The mule deer buck may, however, occasionally strut around with its tail nearly vertical, probably as a sign of confidence. But the smaller black-tailed deer of the Pacific Coast carries its tail nearly level.

The mule deer may form into small bands during the winter (larger bands or herds in the more open desert country), but the blacktail rarely forms bands. The mule deer usually bounds over the ground with great, jolting jumps instead of skimming along close to the surface as does the whitetail. The motion of the blacktail is in between these two. But all three kinds of deer drop and quiver the tail when they are hit badly by a hunter's bullet.

In the early fall hunting season in the West, both black-tailed and mule deer bucks go through a baptism of fire that sees many of the youthful and inexperienced bucks killed. The wise bucks hide in the deepest thickets and learn to run low to the ground at this season, instead of with their usual great jumps. They also have learned to hold perfectly still and give no alarm cry.

The mating time in the late fall sees the bucks uttering barking

Vegetarian Mammals

challenges to one another, and doing much snorting and stamping of feet. The does are rather silent creatures, answering their mates mainly by movements, such as twitching ears and tail. The tail pressed down usually means the doe is ready to mate, and both does and bucks play chasing games at this time, round and round through the bushes and trees. As the buck chases after the doe, he may give a low and deep baa of warning to the fawns, who try to follow the mother, to get out of his way. If they do not, he may butt them, not too gently, with his horns. Fights between bucks are similar to those of the white-tailed deer.

The young are generally born in May and stay quietly in hidden places until they can run with their mothers. I stumbled on such a fawn once in the California mountains, and it gave a quivering bleat of surprise and terror. The first cry of the very young fawn is like the squeak of a rat. This gradually grows into a stronger bleat, a cry for food or protection. When the mother comes running, she bleats an answer, and her rump patch bristles if she senses danger. If there is no danger, the doe gives a soft, murmuring sound of reassurance as she comes near the place where the fawn is hidden.

Each doe usually claims a territory at this time of year and rushes

Mule deer: buck, doe, and fawn (concealed in bushes).

at other does to butt and kick them if she thinks they are entering her territory. She bristles, and her ears are laid back as she attacks. The bucks are off by themselves at this time, in small groups, and seem to show much less interest in territorial claims.

During part of the year, particularly the winter and early spring, the mule deer of the more open country, such as the sagebrush desert, form into bands that are usually led by a wise old doe. If she sees something strange ahead, as she leads the way along a trail, she may snort or blow with curiosity and stamp her feet. This noise is partly to make the strange object or thing show whether it is alive or dead. All the other deer become alert at the sound. If the lead doe thinks it is dangerous, she gives a high-pitched whistle or snort that sends the whole herd pounding away in great leaps for cover. Mule deer of the woods and black-tailed deer have much smaller bands, often just one family.

Each mule deer has a scent gland on each foot and in the hock that it apparently uses for communication. These glands are rubbed against trees, and the smell may be used to lead other deer to good pasture or along safe trails to water. A buck may use the strong smell of his glands to frighten away smaller bucks.

Recent studies of the black-tailed deer show that this animal has at least six separate scent glands and pheromones. The deer anoint their foreheads with pheromone scent from their hind legs. Their metatarsal and interdigital glands mark the ground with scent, the first as the deer lies on the ground, and the second as it travels. These special scents are used to relay messages to other deer. Usually the ground scent simply means "I have been here. You can follow where I am by following this scent." Other scents mean, in the case of a female, "I am ready to mate," or, for a male, "I am seeking a mate!" while still another scent means "Danger is near."

Elk

The elk, or American wapiti, is a large deer that has developed the harem system more completely than any other American deer. In the late fall, each of the great bulls gathers a harem of cows, sometimes numbering more than twenty, and guards them from all comers. The young bull must break into this system by stealing one or two cows

Elk bugling.

from an older bull. Sometimes he is caught and forced to fight, and may be killed. The wisest of the young bulls go off by themselves to feed and strengthen themselves and wait until they are very large and powerful before they tackle the harem masters. To go off alone in this way is said to be a sign of future greatness in a young bull.

The magnificent bugling cry of the bull elk rings through western mountains in late October. The bull gushes forth with a vast and guttural roaring that rises in pitch to the shrill tones of a trumpet, and finally reaches a whistle that is like a harsh, jarring scream. This fades and drops back again to deep, guttural sounds that end with a series of fierce grunts. Only the great bulls in their prime can give the complete bugle. Younger bulls, because of their shriller tones, are called "squealers."

As two bulls cross a clearing to meet, the world seems to come to a standstill to watch. The manes of the bulls rise and their swollen neck muscles bunch and throb. They snort loudly, paw the ground, and shake their antlers at each other, then rush together with a jolting crash. The fight may go on for a long time of straining and pushing

Elk bulls fighting.

and pulling, of breaking away and crashing together again, or it may end quickly with a sudden goring of the weaker bull or with one turning to run.

During most of the year, an elk herd is led by a wise old cow, while the bulls follow meekly behind or go off by themselves. I have passed near such a herd in Yellowstone Park and heard them noisily conversing as they ate, with a series of squeals, bleats, and barks. Mothers bleated for their young not to stray too far away. Young does and bulls squealed with delight or half rage as they chased each other, while an occasional barking cry might have been a call to come to better food. Another and more commanding bark from the leading doe meant something like "Cut the foolishness and follow me!"

The bull is not the only one to bugle, as the mother elk may give a triumphant bugling cry after the birth of a fine calf. This is soon answered by the high-pitched bleat from the calf, demanding milk. A mother elk has been heard to give a sharp bark of warning and distress when a grizzly bear approached her young one. Other elk may come to help.

In times of stress or excitement and as a warning, the hairs of the rump patch of adult elk are raised and displayed in a ripple of lighter color, as in most deer. At such a sign from a neighboring elk, other elk may turn and run.

In summer the bulls leave the main herds and move to the highest pastures in the mountains, where they can avoid flies and find safe places to grow their velvety horns. In late August such bulls have been seen to take part in a most remarkable dance. They move about ponderously in a circle with heads low to the ground, hoofs stamping and pawing up clouds of dust. The most likely purpose of this dance is simply to drive away flies and other insects that attack at this time of year, but it may also be a social dance.

Moose

Moose are the giants of the deer world and have been known to stand up to a grizzly bear in fair fight. Moose do not develop harems in the way of the elk but may visit several cows during the mating season.

The bull begins by mid-September to look for cows and continues the search for about a month. He searches particularly for open glades of hardwoods along the ridges, and in these he searches, with nose to the ground or lifted high to drink in the wind, for some hint of a cow. His great ears flap back and forth listening, and every now and then he gives a deep grunt of "Oo-wah, Oo-wah!" or a short bellow. If he hears the long, mellow, and hankering "Er-er-er-er!" of an eager cow, he goes crashing through the trees and brush as if they were not there. The cow calls the bull not only with a gentle bellow but also with soft squeals and whines. She may run a short way but soon waits for him. When the cow and bull meet, there is much grunting and mumbling, and much nosing of tail and head and rubbing of sides.

If the bull, instead of hearing a cow, hears the challenging grunt or bellow of another bull, he also rushes through the trees, but this time he makes a great noise of slashing the brush and branches with his widespread antlers, hoping to impress the other bull with his great size. He circles to catch the wind from the enemy and bellows again and again. When the bulls get near each other, they approach more slowly and, with much grunting and stamping, try to maneuver for the best place to fight. The hair rises stiffly along their shoulders and neck, and they seem to grow in size and strength. Again and again the brush is slashed with the heavy antlers. Then suddenly they charge together head on, their horns acting as both spears and shields. The stronger bull finally chases the other away.

The moose calves are generally born in May, when the cow seeks a secluded island or swamp where she can feel safe from enemies. In a thicket of brush the mother hides her baby for a week or so. She soon brings her calf out to show the world, but when danger comes, she gives a warning squeal that sends it running to hide. Then she lifts her head, pulls back her ears, snorts like a vicious horse, and bounds forward on her long legs, striking the ground savagely with her forefeet, her eyes red with rage. Bulls have been known to give a warning grunt that sends both cow and calf running for cover. The calf bawls for help something like a lamb, with a loud "Baa-baa-baa!"

Like the deer, bull moose make soiling pits or wallows in the fall, and the bad scent of these no doubt helps spread moose talk through the woods, telling of the biggest bulls. Moose make beaten-out yards in the snow in the winter to give them running space against enemies like the wolves, and here they often depend on the shrill cries of jays to warn them of danger.

The flexible nose of the moose expresses many emotions, such as fear, anger, disgust, and love. I was sure that the big bull moose who once frightened a group of us into our car with his foot-stamping wrinkled his big nose and opened his mouth in a soundless laugh at the humans he had scared. Fear is shown by the narrowing of the nostrils, anger by the nostrils flaring outward, and disgust by repeated sniffs and shaking.

PRONGHORN ANTELOPE

The pronghorn antelope is a creature of the open American plains, where its tremendous speed and far-seeing eyes help preserve it from its enemies. It is more closely related to the American mountain goat than to the antelopes of Africa. Pronghorn are great racers and hate to have anything beat them. If a man on horseback races with a pronghorn, the pronghorn is delighted and tears off parallel with the horseman, keeping up about the same speed and toying with him for a minute or two. Then he suddenly puts on a tremendous burst, forges way ahead, and cuts across in front. A magnificent male pronghorn once tried this when my wife and I were driving a car over the Colorado plains one day. At forty-five miles an hour we raced parallel with him for a full minute; then he put on a great effort and moved out ahead of us to swing over and cut across the road in front of us. Despite the rough road, I pushed down the accelerator and brought the car up even with him again, both of us going a full fifty miles an hour. The buck was running with neck stretched out and his legs pounding so fast they were only a blur, but he could not pass us.

Pronghorn have remarkable control over the hair on their body and can use it for signaling. The mane on the neck, especially a buck's, bristles erect when he is excited and angry, acting as a threat and bluff to another buck. But the white area on each hip is much more effective as a signaling device. The hair on this area is specialized for signaling; short in the center, it gradually becomes longer toward the front edges. A special muscle under the skin can be used to make the hairs move any way the pronghorn wants it to. This hair can be suddenly raised and spread in radiating form into two great panels of dazzling white that catch the sun's rays and flash them far across the prairie.

A pronghorn I scared on the Colorado plains suddenly flashed this dazzling white signal, and instantly flashes of similar signals from other antelope in the nearby brush caught my eye. In a few seconds the entire herd had dashed out of sight over a rocky ridge. These signals can be seen from many miles away. To help further in signaling danger, each animal gives out a peculiar musky smell that is carried on the wind and may reach and warn antelope that cannot see the flash of the white rump patch. Both friendly and warning messages may be flashed and scented in these two ways.

Pronghorn have several other smell glands that are probably useful in communication. The glands on the jaw are largest on the buck and grow biggest during the mating season in October, when they probably send forth a special smell message to the females and a warning to other males. The glands on the foot and hock are probably useful in leaving scent messages on rocks and bushes among which the antelope passes. Another antelope coming later can tell from the smell whether a friend or stranger has passed, male or female, frightened by a man or afraid of nothing, and so on.

Young pronghorn are born in the spring. The mother may leave her baby alone for some time while she goes off to feed, but she has a trick that is very effective, and that is to kick up a cloud of dust all around where the little one lies. This dust settles on the antelope kid and completely disguises his smell as well as her tracks. Smart mama!

Pronghorn antelope herd.

The mother calls her young one with a querulous, grunting bleat, and the kid answers with a shriller, softer bleat or squeak.

When it is not sure of what it is watching, a pronghorn may give a short bark of curiosity. I came near a herd once at night and was startled by a combination shrill whistle and snort of alarm that sent the whole band crashing away into the darkness.

Sometimes a mother uses a loud, grunting bleat to lure an enemy away from the neighborhood of her young. If this does not work, she may pretend to be lame and limp away so she will be followed. If the stranger still goes near the baby, she may suddenly advance with bristling hair and blazing eyes. A blow from a sharp hoof can split a coyote's skull.

In the summer the pronghorn does and their young begin to gather into small bands. Later the bucks join them, and at such times the young ones have many merry games of king-of-the-castle, hide-and-seek, and follow-the-leader—bleating and squeaking with delight.

The larger bucks gather harems of does in the fall, and sometimes fight vicious battles. Two advance, bristling and snorting and stamping the ground, and spar at each other with their sharp horns like swordsmen. Often the winner continues to chase and gore the loser until he is dead. A buck can give a loud "Kaaa!" as a cry of triumph.

WILD SHEEP

Bighorn sheep can perform astonishing feats of climbing among the rocky cliffs of the western mountains. Far more intelligent than domesticated sheep, they cleverly avoid the dangers of both men and mountain lions by posting guards to warn of danger. A warning snort sends the whole band rushing up to cliffs and peaks where the enemy cannot follow.

The baa of the mountain sheep is similar to that of the domesticated sheep and means much the same; their actions at mating time are similar. But they have far better eyes, ears, and noses for spotting peril, and when an old ewe throws up her head to watch across a mountain slope for danger, the rest of the band is also instantly on guard. Silently they may melt into the brush and disappear when the leading female gives the commanding "Baa-a."

PECCARIES, OR NATIVE WILD PIGS

The collared peccary of the southwestern United States, from south Texas to south Arizona and most of northern Mexico, is a member of one of the most interesting kinds of animal social groups. This tough-skinned, stiff-haired, and piglike creature is among the bravest and most wily of living creatures. It is true that he usually runs as soon as he sees a human coming, but this is not fear so much as common sense. Humans have guns, and the peccaries know this and respect them. But if you corner a herd or trap an individual, watch out! They fight to the death with sharp tusks. They are also very clever, for they know exactly how to get into the thickest brush possible in the shortest amount of time when a dangerous enemy is after them.

Herds of from two to twenty individuals are common in brushy and hilly country. They usually forage in the early morning and in the evening, and find thick brush to sleep in at other times; the herd always has one or two guards constantly on the alert. Their squealing, piglike grunts sound instant warning of danger and a deeper grunt calls the herd to battle. Herds often eat under nut trees, such as oaks, for their favorite foods, and at these times they are in most danger from a mountain lion or jaguar jumping down to grab a young peccary that has wandered away from the herd. But the scream of a pig who is in danger or attacked brings the herd to its rescue within seconds. Even a two-hundred-and-fifty-pound jaguar who has grabbed a young peccary is in trouble when the herd war cry sounds and the forty- to fifty-pound wild pigs come on the scene, utterly fearless and savagely angry.

RABBITS AND HARES

One difference between a rabbit and a hare is that rabbit babies are born pink and hairless, while the hare has furry babies. The rabbit also runs into cover or underground when chased, whereas the hare almost always depends on its long hind legs to get it away from danger. For these two reasons a jackrabbit is actually a hare.

The language of rabbits and hares is not extensive, as most of them are not highly social and they rarely cooperate with each other. The sounds are similar to those of tame rabbits, including loud squeals.

Rabbits

The cottontail is found almost everywhere in the United States and southern Canada, even in some parts of the deserts. Nearly everyone has seen the little, dodging streak of the cottontail as it rushes for home or shelter when surprised, its white tail standing up on end and flashing a signal of warning to all nearby rabbits.

At mating time, the male and female, buck and doe, may hop over each other in a kind of dance and rub noses and bodies together. The male often nuzzles and licks the female's fur.

A cottontail doe digs a shallow burrow under a log or rock or in a dense thicket and lines her nest with fur from her own body. The young rabbits soon after birth begin to squeak faintly for food. Later they are able to squeal when in danger, and sometimes the mother whines to her young. She protects them by kicking the enemy with her long hind feet.

On moonlit nights cottontails love to gather in a clearing in the woods and dance and play, sometimes jumping around and over each other as if they had gone crazy. But a sudden thumping from the hind feet of a sentinel sends every rabbit dashing for a hole or thicket.

Mother rabbits lay claim to definite territories in which to raise their young, and they fight all rivals. When two females fight, they advance on each other with grumbling noises, their ears flat down; then they box with the front feet and try to bite. If a bite succeeds in getting a grip, each animal curves her body toward the other and kicks with her hind feet. Males fight in the same way.

Hares

The two main kinds of hares, or jackrabbits, are the blacktail of the more southern and dryer, lower country, and the whitetail of the more northern and mountainous country, though the ranges overlap in the West.

The whitetail, a larger and stockier hare, flashes its white tail when surprised, both to confuse an enemy and to warn other jacks. In the far Southwest, the antelope jackrabbit, a variety of the blacktail, can move the white hairs on its rump exactly like the pronghorn, and so flash warning signals to its friends as it runs. It also seems to flash the dazzling whiteness directly into the face of a chasing enemy, partially blinding or befuddling it.

The jackrabbit at first freezes perfectly still when an enemy approaches, hoping the hunter will pass on without seeing it. If discovered, it dashes away in a zigzag at high speed, often making two or three particularly high jumps so it can look back.

In the mating season, which lasts, off and on, from early spring to midsummer, several males may gather about one female. They chase each other, spar and box, and sometimes come to close quarters to bite and kick. There is some angry grunting and growling in such combats. A young jackrabbit may both spit and box when attacked.

The snowshoe rabbit, or varying hare, of the Northeast and the high mountains of the West, is really a hare and, like all hares, depends mainly on speed to escape its enemies. It also has wide, flat feet, which help it run over the surface of soft snow. The males have many boxing matches during the spring and summer mating seasons, and sometimes they give snorting grunts of anger. The female usually runs away from the male and leads him a merry chase in and out of the bushes and trees. Most of this hare's methods of communication are similar to those described for jackrabbits.

RODENTS

Rodents have four large front teeth instead of the overlapping eight of rabbits. They are the most numerous of all mammals in kind and number, and form the staff of life of many birds and carnivores. Because of the very many kinds of rodents, I am going to discuss the languages of only a few of the more well-known species. Little is known about many species of rodents, and much could be found out about them by careful study.

Native Mice and Rats

White-footed Mice. The most typical native mouse, and the most successful, is the white-footed mouse, which is found nearly everywhere on the American continent, from low desert to high mountain peak.

Like that of most mice, its chief defense against a hostile world is to breed as many children as possible.

One of the most interesting of all my wildlife experiences was to sit very quietly in an old, abandoned cabin and watch the white-footed mice make merry among the decaying quilts and old newspapers. These mice, as I became more acquainted with them, seemed most astonishingly human. Old and young played together like children, one sometimes sneaking up on another and pinching its tail, then both chasing each other round and round an old oilcan at top speed, then reversing and chasing the other way.

Every few minutes a mouse would stop its play or its search for food to clean itself. It did this seriously, by first running its tail through its mouth, then reaching down and grooming all the fur it could reach. It would even comb and slick down the fur on top of its head and neck by moistening its front paws and running them over its "hair," like a man standing before a mirror. Last, with a pleased expression, it would look itself all over and then run back to play.

These mice seemed to communicate whenever two of them met. I suspect some of this language lay in the expression of their large, liquid, brown eyes. Sometimes they would wiggle noses together or sniff each other all over. When a mother met one of her young ones, you could tell it was hers because she would stop to lick it. Once in a while two mice would rear on their hind feet and playfully box with their forefeet. But once, when two males who did not like each other met, they squeaked angrily, rose, and boxed vigorously, then clinched and rolled over and over biting at each other. The smaller one finally tore himself loose from the other's grip to run away, squeaking in loud protest. The winner of this particular fight seemed to be the boss of the cabin, for I noticed that usually whenever he came up on stiff legs and with bristling fur to another male, the other ran.

When I moved a little bit, a mouse that saw me immediately thumped the floor rapidly with its forelegs, and I at once heard, all about me, the scampering of little feet running for cover. However, I held completely still. Soon I saw a tiny nose peek out from a hole in the wall; then there followed a squeak of reassurance, and several mice came out to play.

White-footed mice, as well as most other mice, develop definite territories that they defend. Most white-footed mothers raise their young alone, without the help of the father, but in one species this is not true, and the father becomes a true provider and helpmate. If a strange mouse comes near the territory of the family, he is warned away with angry squeaks and with thumpings of the front feet. If the hair rises on the back, and the ears lie back, this is a warning signal of imminent attack.

The many different species of white-footed mice show a very wide variety of behavior. Some are very friendly and make easy pets (most of the western species), while others (especially some eastern mice) are very nervous and short-tempered.

The female raises several litters a year, usually in some hidden place in a bush or under a rock or in an abandoned building. The mother answers her babies' high squeaks of hunger with reassuring deeper squeaks of her own. She hunts for food mainly in the early evening and very early morning, sleeping with and nursing her young ones the rest of the day. Though seeds and plant foods of various kinds are mostly sought, she often creeps up on insects like a little cat and suddenly pounces. Even other smaller mice are sometimes caught

Mice: (clockwise from upper left) *a female mouse in the nest with her babies, the male looking on; a mouse cleaning itself; mice preparing to fight; mice disputing over a nut; a grasshopper mouse screeching; a mouse catching a bug.*

and eaten. As she hunts, she is constantly on guard against danger, her nose wrinkling against every breeze, her whiskers quivering and her ears and eyes alert. At any dangerous sight, sound, or smell, she runs with a squeak of warning for a hole.

Occasionally a white-footed mouse gives a shrill, buzzing sound that may be heard forty to fifty feet away. This is probably the mating song of the male. When he finds a female, he chases her but may suddenly seem to become frightened and run away from her. Then she chases him. The two often smell and sniff each other all over with great excitement. Field mice share this behavior.

Grasshopper Mice. Mention should be made of the howl of the grasshopper mouse, a stout little mouse with a naked, thick tail. This mouse has apparently become a carnivore, for it lies in wait for and pounces upon other mice and large insects with all the fierce savagery of a tiger. Its faint, shrill, and long whistle has been likened to the howl of a wolf by many naturalists, because the mouse gives this "howl" whenever it starts out on an evening or early morning hunt. One purpose of the howl may be to warn other grasshopper mice to stay out of the howler's territory.

When following down a hot trail through the thick grass jungle, the grasshopper mouse may give a sharp, squeaking bark of excitement. Possibly this sound is used, like the roar of a tiger, to paralyze the will to resist of the mouse who is about to be attacked.

Wood Rats. The wood rat is a far more likable and beautiful animal than the Norway rat that so infests our cities. I have lain out in the woods and watched a wood rat come out of its big pile of sticks to search for food, its whiskers quivering in every direction and its nose wrinkling to catch the scents. The large, brown, and liquid eyes appear very alert and intelligent. The body is pleasantly clothed in clean brown fur above, white and whitish below; the tail also is covered with hair, far different from the ugly, brown, and dirty common rat with its cunning and savage small eyes.

Wood rats are not very sociable animals, though individuals vary in this regard. Each individual either occupies an old stick house or builds his own new one, and tries to find a place that is best protected from the rain and from enemies. The best houses are occupied by the biggest and strongest rats, who drive away all others of their kind. As the rat sits in his house and hears another rat approaching, his immediate warning signal is to thump his tail vigorously on the ground. If this fails to drive the intruder away, he gives an angry squeak and rushes out to do battle. Usually the strange rat makes a hasty retreat.

The squeak of the rat conveys different emotions, such as anger, or fear, or the calling for a mate. A wood rat I caught in my gloved hands once squeaked very fast and loud at first, but the sounds gradually became softer.

The wood rat gives many signals by its movements. When the ears are held back and together and spread to catch sounds from the sides, the rat is listening very carefully for danger. As soon as it senses an enemy, it loudly rattles its tail in warning and dives into a hole, where it rattles again. If the large, dark eyes bulge outward, the rat is feeling great fear. If the head is thrown up quickly as the animal is feeding, it is alarmed. When the ears flare up and outward and are rigidly held like cups, the rat is curious about some noise it hears and possibly a little alarmed. Ears completely flattened back and whiskers quivering mean anger toward another rat.

The wood rat's house also communicates to other rats and to us. If the entranceways at the base of the stick pile are well cleared and show fresh droppings nearby, we know there is a rat using the house. If it is a low and poorly built house, it is probably being used by a young and lazy rat. If the owner doesn't get busy and built it right, he is liable to be either washed out or caught by a fox or an owl. When a house is well built, in a good location on a hillside and with a steep roof, it is probably occupied by a wise old rat. A rat seen taking soft nesting material into a house is a female preparing a nest for her babies.

Rats leave their smells about their homes and in nearby places where they stop to wet, leave droppings, or gnaw on twigs. A strange rat visits all of these places, sniffing deeply in order to determine what kind of rat left the signs. In the winter and early spring, the males are looking for mates, and they approach these smell posts with great excitement, their whiskers quivering.

When two hostile males approach each other, they rattle their tails and stick out their hair to try to look as large as possible. They squeal angrily, chatter their teeth, and will fight unless one is bluffed away

Vegetarian Mammals

by the other. Tails are held high until the bluffed rat drops his and runs.

You can tell a male at mating time by the way the heavy cheek muscles give its face a puffy appearance. The belly is often stained brown by a smelly secretion. This stuff is rubbed off on rocks or twigs to leave messages to females and warnings to other males. A large and powerful male may have his house surrounded by houses of females, since he drives all other males away from the area near his house. After the female is mated, she soon drives the male out of her house with angry tail rattlings and bites.

Low chirps have been heard from wood rats during the mating season. Certain wood rats not only rattle their tails when alarmed but also thump the ground with their hind feet.

Wood rats are also called "pack rats" and "trade rats," because of their habit of running away with any bright and shiny objects they may find. I once watched a wood rat that was carrying a stick immediately drop the stick when it saw a bright sixpenny nail. Its whiskers quivered with excitement and its eyes shone with avid interest as it

Kangaroo rats fighting.

picked up the nail and ran away to store it with other bright and shiny things in its home.

Kangaroo Rats. The kangaroo rats of our deserts and drier country are among the most interesting of all American animals. Their powerful hind legs, long, graceful and tufted tails, and cleanly habits make them distinctive. The males fight by leaping high in the air and kicking at each other with those strong hind legs until one is knocked unconscious or severely wounded.

Kangaroo rats vary greatly in size and character among the many American species. Some are gentle, confiding little creatures who make excellent pets, while others are savage and untamable. In the wild, one species of large kangaroo rat may completely dominate a smaller species, which it chases away from the best food and kills if it captures it.

Kangaroo rats live in holes under bushes and come out only on dark nights to feed and fight and play. In the desert darkness I have heard the low chuckling noises of food-hunting rats, heard the high-pitched squeak of anger when two fighting males meet, and listened to the loud thump of warning given by their hind feet when they feel danger is near. But the rats themselves I could see only rarely as dim and shadowy elves in the starlight.

Beavers

The beaver, though the most intelligent and highly social of all rodents, is by no means as clever and wonderful as some popular writers have made it out to be. Beavers do not always know how to fell trees in the best direction, and the dams they build, while showing great industry and construction ability, are not always made in the best places or with the best design.

Beavers generally mate in January or February and may mate for life, though some males appear to break this rule. Males, seeking a mate, leave scent from their castoreum gland in mud pies along the banks of creeks or ponds, and also search for and smell excitedly the scent left by females. Males bristle and chatter their teeth when they find the scent of another male.

The kits, or baby beavers, are usually born in May, and soon grow large enough to start helping their parents with their work. The summer and fall is a very busy time, spent in building and strengthening houses and dams, storing food for the winter by hauling branches under water and anchoring them with mud, and extending their canals to new feeding places.

I watched a beaver come out of the water once and stand erect for some time, testing the air with its nose and turning its ears from side to side. Satisfied that no one was near, it waddled to a small cottonwood tree up on the bank where it had been working before, placed itself before the tree with legs spread wide and its thick, flat tail forming one arm of a tripod. Then it seized the tree with its front paws and began efficiently to cut out large chips with its strong front teeth. It had hardly started, however, before another beaver saw a slight movement that I had made with my hand and hit the beaver pond a terrific slap with its tail. Instantly the beaver on land turned and darted for the water, into which it dived and disappeared.

As a beaver gives warning of danger, there are actually two sounds, first the loud slap of the tail on the water, then a hollow plunging sound as it dives. The second sound confirms the first. Beavers also

Beaver: (clockwise from upper left) *beavers diving; beaver mother and baby; a beaver in a tunnel.*

have several other noises of communication, including a hiss of menace when ready to fight, a chattering of the teeth in anger, a less bellicose chattering noise when they meet in friendship and nibble at each other's cheeks in greeting, a querulous "Chrrrr?" when they are annoyed or questioning, a childlike wail from the young when they are hungry, and a deep, hoarse groan from the male when he is calling the female at mating time. Rarely, beavers will snarl, when engaged in play fighting or the real thing. Soft churring noises, mumbles, and whines have been heard from inside a beaver house and these are probably social noises or requests for more room and food.

Beavers are very cooperative in building dams, stick houses, and canals, so that the feeling comes to many who watch them that an old and experienced beaver carefully plans and directs all their work. But the actual communication between them that tells each what to do has not been observed. I have watched beavers rub noses, softly bite cheeks, and mutter together, so that some message about work could have passed between them. I am inclined to believe, however, that the old leader mainly shows the others what to do by example and that they copy him. Possibly he comes around now and then to check on their work and criticize or encourage, largely by his attitude. Animals are extraordinarily keen about recognizing emotional meanings from the tiny movements of other animals or the expression in their eyes.

Squirrels

Black-tailed Prairie Dogs. Prairie dogs are closely related to the ground squirrels. At one time, prairie dogs had towns all over the Great Plains, but ranchers wanted the ground for grazing cattle, and did not want their horses to break their ankles in prairie dog holes. But now there are some protected prairie dog towns in national parks and monuments and even on some ranches, where small towns have been preserved for their scenic charm.

These thirteen- to fifteen-inch-long animals, yellowish-brown in color but with black-tipped tails, are great fun to watch. They live in towns where their cooperation is mostly seen in their whistles of alarm when danger is sighted and all prairie dogs take to their holes. If you sit very still for a while, they will soon come out of the ground to play or to forage for the plant food they crave. They talk to each other with body contact and movements as well as with noise, two animals often rubbing bodies together to soothe itches or just for social contact, but they also chatter loudly at each other when they do not want body contact. The most spectacular behavior of these animals is the jump-yip display, in which an individual jumps off the ground and makes a loud "Yip" sound. This display is usually given by an individual who has been suddenly startled in the process of defending its part of a prairie dog town, although sometimes it is a call to dust-bathing, foraging, or other activity.

Gray Squirrels. The first time I heard the loud "Quack-quack!" of warning given by a gray squirrel, I thought the noise came from some big animal like an otter, and rushed down to the Eel River to look. Soon I saw the gray squirrel in a tree, jerking its large and graceful tail about, and heard the startling noise again. The signal was followed by complete silence, as every gray squirrel in the neighborhood had found a good hiding place and was staying there.

Other sounds given by the gray squirrel are a peculiarly defiant "N-grrrr, n-grrr, n-grrrr!" when the squirrel feels it is successfully escaping or hiding from an enemy; a low purr of pleasure, when one squirrel is being scratched or stroked by another; also a "Maek-

Black-tailed prairie dogs perform the jump-yip display.

maek!" with or before the cry of warning, expressing intense excitement, fear, and defiance; a scolding "Kwa-kwa-kwa!" used to drive jays or enemy squirrels away; and a nasal, throaty, buzzing series of sharp grunts, which is sometimes heard to come from two or more localities at once and usually expresses triumph. When this last sound is soft and low, it is like a purr and expresses love between mates. If it becomes very violent and loud, it is an angry rebuke and may be a warning before one male attacks another. Another similar warning is a chatter of the teeth that sounds like a muffled clicking of rocks in a box.

The barking song of the squirrel is heard sometimes in the early morning and may be either the call for a mate or simply an expression of delight in being alive. It may be first a series of sharp, rhythmical cries like a cat crying, then a rolling string of "Qua!"s that gradually quickens into a harsh, rattling drumbeat. The song varies, however. When it is a real love song, either male or female will put into the call a longing and tenderness, so soft at times as to sound like the whimper of a baby mouse or the hiss of trees rubbing together in the wind. The squirrel, as it sings, stands with body tense, tail quivering, and throat throbbing.

Red Squirrels. The red squirrel is a creature of the pine forests, unlike the gray squirrel, who prefers the hardwood forests. The constant chattering, chirring, coughing, barking, and scolding of the red squirrel indicates its high spirits and mischievous nature. The tail is a blur of motion, expressing at one minute curiosity, at another anger, and a third time mocking triumph. The forefeet may be stamped in anger or as an alarm signal.

Red squirrels sometimes combine all their various cries—the coughing challenge, the spitting hiss of anger, the "Chirr-chirr" cry of comradeship, the growls of rage, the "Snick-snick-snicker" scolding cry—into one song that becomes a song either of love or simply of good spirits. Red squirrels commonly mate for life; there is often real affection shown between the mates as they nuzzle each other and chirr softly.

Both red and gray squirrels have glands in their rear ends that give out a clear, yellow fluid or musk that is dropped here and there on the trunks of trees. These scent posts or signal boxes are smelled carefully by other squirrels and pass along useful messages.

Squirrels in various aspects.

Ground Squirrels. There are so many ground squirrels that it is possible to make only a brief mention of a few outstanding species. All ground squirrels have thinner tails than the graceful-tailed tree squirrels, and when in trees, they react to danger by running to the ground.

On our ranch, the California ground squirrels whistle shrilly whenever they see us coming, and dart for their holes. Some of the more cheeky stay at their burrow entrances, jerking their tails rapidly, and calling us names with each sharp whistle.

In the central Rockies, the Columbian ground squirrel gathers food in the fields whenever the sun shines, chattering in low chirps with its neighbors. In the mountains of the West, the golden-mantled ground squirrel is commonly found on the edges of the pine forests, where it is often mistaken for a chipmunk because of its gold body stripes, although it has no stripes on its head. It occasionally gives a single alarm chirp that sends its friends into their holes. "Tachack-pr'r'r'r!" is another alarm note given, after which the tail is jerked violently. When angry or impatient with other squirrels, it grunts, buzzes, and chirps at them. Sometimes a soft tick-tick noise, with nervous jerks of the tail, expresses worry or dislike.

The antelope ground squirrel of the southwestern deserts has a white-marked tail that it flattens over its back and twitches quickly or even vibrates with great speed to flash a message of alarm. Sometimes it gives a shrill, rapid chittering cry when alarmed. The rock squirrel of the southwestern mountains spreads the alarm with a short, sharp whistle. The striped ground squirrel of the Great Plains also has a short, sharp whistle for spreading alarm; it also gives a long-drawn-out and quavering whistle of defiance when it escapes to its hole, or an angry snarling when trapped or fighting with its kind. It also snarls when attacking a mouse.

Woodchucks. The woodchuck is a kind of large ground squirrel found mainly in the East. The yellow-haired marmots of the western mountains are close relatives. The hoary marmots of the highest parts of the mountains are the largest, and have the strongest voices. All three animals are called "whistlers," because of their loud whistles of alarm. Small red musk glands on the inner sides of the front legs of the woodchuck are used to leave smell messages.

Squirrels and chipmunks do a good deal of talking with their tails, constantly jerking them about by nervous action of the muscles at the base of the spine. When a squirrel feels real fear, it lowers its tail and runs for dear life. When it sees a strange creature or man whom it wants to scold, the tail begins to jerk about furiously while it calls the other creature names in a loud chatter.

When one male squirrel is advancing to attack another male, he spreads the hairs on his tail to make it look as large as possible and holds it high and curved over his back. If a tail starts to lower, one squirrel is becoming afraid of the other and getting ready to run.

Porcupines

All the porcupines I have approached have gone through the same warning ritual. First the porcupine rattles his tail at me, then he begins to bring all his quills erect. Third, he humps up the middle of his back, pulls his nose and head back, and raises his tail in a position for immediate action. All this means "Look out! Don't get too close!" When I touch the porcupine with a stick, he lashes suddenly at it with his tail, trying to put quills in it, and at the same time chatters his teeth in anger.

When I try to circle a porcupine, it turns its body with me, moving the hind quarters with quick little hops of its hind legs so that the dangerous tail will always be toward me. This technique is very effective with dogs, and the attacking dog usually comes yelping home with his mouth, lips, and neck full of quills, each of which continues to eat its way into the flesh.

During the mating season, male porcupines bristle their quills at each other and chatter their teeth in rage before attacking. All porcupines at this time become very loud-mouthed—grunting, whining, chattering, or barking and mewing at each other. The hankering notes are the talk of two mates, the angry ones a dispute between two males. Very ardent males may moan.

BATS AND INSECTIVORES

These animals are discussed together here because they are small and are not noted for very elaborate forms of communication.

Bats

Much of the communication of bats is lost to us because their squeaks, for the most part, are so high-pitched that we cannot hear them. Most of them do have squeaks of alarm that we can hear, however, and in a cave where there are thousands of bats, my ears have been almost deafened by these shrill sounds. The extraordinarily quick movements of bats when flying also must have value to the animal in communicating, telling other bats of the location of their flying insect food and also of the direction of danger.

Insectivores (Moles and Shrews)

These very primitive animals lead lives that are largely instinctive. The shrew, for example, travels at high speed over the ground in a zigzag fashion, following its own smell trail or that of other shrews, and reacts to the smell of a mouse or insect by whirling about until it can make contact and kill. But, put in a strange position, it appears helpless and unable to know what to do.

The underground life of moles is yet to be really explored by naturalists, and little is known about their language. I have kept shrews in captivity and listened to their high-pitched twittering, a sound that becomes very rapid and bloodthirsty when they start to fight. Females make a low-pitched, rapid, chattering noise when nuzzling a male mouth to mouth in a sign of love. A very shrill chatter or shriek is given by a trapped shrew. A very soft squeaking sound is heard when a shrew is contented and happy.

Shrews give out a strong smell from glands at the rear of the body, and leave it wherever they go, not only as a message to other shrews, but to remind themselves where it is safe to travel.

ARMADILLOS

The armadillo is one of the strangest creatures in America. It is the only animal that is heavily armored all over its body except for the underside. The creature most like it is the porcupine, which has sharp quills instead of armor all over the upper part of the body. While the porcupine is a tree-climbing animal in the heavily wooded areas of the north, the armadillo is a ground animal native to Texas and northeast Mexico. The northward spreading of the armadillo into Oklahoma and beyond is limited by the fact that it cannot live in cold weather for long; the armor on its back does not keep it warm, and because of its very poor, peglike teeth, it must have access year-round to insects, worms, and other easily chewed things that live in ground that does not freeze.

The armadillo communicates, usually between the sexes (it is not social otherwise), by soft grunting noises and by its actions when attacked. It either runs as fast as it can over the ground to get into a hole, relying on its back and head armor to protect it from a wildcat, dog, or other predator, or, if too far from a hole, it may roll itself into a ball that is protected on all sides. In both cases, it simply says, "I am too hard! Leave me alone!"

The mumbling and grumbling of two armadillos that come together is either preparation for mating or the attempt of one male to drive another away, which he cannot do by force, as armadillo armor is far to good too allow them to fight.

6
Wild Birds

The language of birds, with the exception of a few highly social and intelligent kinds like the crows and the parrots, is generally largely instinctive and is simpler than that of mammals. For this reason, I will begin my discussion with certain basic information about bird languages plus some examples of a few birds that are of special interest or representative of certain orders or families. This information will allow you to start your own studies of specific bird languages.

BIRD LANGUAGE

As with mammals, the main secret of understanding the language of birds is to feel enough kinship with them to catch their feelings and emotions. There is a primitive sense of language in most birds. It is clear that some birds express a few of the more primary emotions. When you hear a gull scream, try to look at things through a gull's eyes, eyes backed by a brain capable of emotion and a swift reaction to happenings immediately around it, but with very little reasoning power, foresight, or capacity for ideas. The gull screams because it is hungry, or frightened, or frustrated, or angry at another gull, or maybe just because it has seen something strange and wants to draw the attention of other gulls to the curious object. Look where the gull is looking, and try to see what it sees; then, perhaps, you can understand the emotion it is conveying with its voice and its motions in the air.

Alarm and distress notes are expressed by a great variety of species. Noisy and alert birds like the yellowlegs, the kingfisher, and the jay act to warn of the approach of danger and communicate not just with other birds but even with mammals. Distinguishing between dangerous and harmless birds or animals, an outstanding capacity of birds, is undoubtedly due to their great acuity of vision. Birds usually signal the presence of a hawk to the human watcher before he has gotten around to seeing the hawk for himself. Oddly, birds can discover a dangerous hawk at an extraordinary distance, but they do not seem to care or notice whether the bird is alive or stuffed and mounted on a perch. Jays will mob a stuffed owl indefinitely.

Many birds use sounds other than singing to attract their mates. A strange humming or pulsating beat made by a snipe to woo its mate is made entirely by the feathers rubbing together. The male ruffed grouse uses his wings to make a loud thundering noise that can be heard a long way; it may also be ventriloquial, so you cannot be sure where he is. His prospective mate is lured to this "thunder." Most astounding of all is the sudden explosion the grouse produces when it rises from the ground; it is probably intended to repel an attack by a predator.

The woodcock produces a shrill twittering that is made by the wings scraping together in some mysterious way. The whistling flight of some ducks, especially the male golden eye, sometimes signals to a female and, in other cases, is a warning of danger. One of the most astounding noises made by wings is produced by the male nighthawk. Once I was made to nearly jump out of my skin when one of these birds zoomed down from the sky and right above my head let loose an extraordinary booming noise that made me think I was being shot at. The male was simply wooing the female.

Some birds have feathers that contain no real color, but somehow act as prisms, suddenly catching the light of the sun and producing hues that are far more brilliant than the regular colors of feathers. These gleaming wonders are produced by hummingbird males when they dive. In some cases, such as in some doves, a film of oil in the wings produces this iridescence. These are probably mainly courtship displays but may also be used to startle an enemy.

Wild Birds

Events that Influence Language

There are usually seven big events in the life of a bird that influence what it is going to say and how it will say it, and most birds react in the same or very similar ways to these events.

The Newly Hatched Birds. The first event is the breaking open of the eggshell and the beginning of hunger. Instinctively, the young bird opens its mouth, and into it the mother or father puts some food that has been swallowed, partly digested, and then thrown up again (regurgitated). The taste of this food is good to the young bird, and it opens its beak and gives the hunger cry of its species, commonly a

A female sparrow feeds her young.

shrill "Tseep-tseep-tseep!" It learns gradually not to give this hunger cry all the time, but only when it sees the parent bird approaching.

Learning to Fly. Actually the young bird instinctively knows how to fly, but at first it is afraid. The parents try to encourage it by calling to it gently as they flutter nearby in the air, by dangling food near enough for it to see but not to reach without flying, and even by pushing the young bird until it falls out of the nest and has to fly. The cries of the parent birds usually become rather shrill and impatient as they get tired of waiting for the young one to make up its mind. "Come on, come on!" they seem to say. "Quit fooling around and fly! You are a big boy now!"

At last the young bird, peeping its protest and fear, is bullied or urged into making the first desperate attempt, which almost immediately shows him he really can use his wings. His flight is very clumsy at first, and he may fall to the ground, but practice makes perfect.

A female Tennessee warbler prepares her chicks to fly.

Encountering an Enemy. The third event, which may be experienced even before he learns to fly and which helps him with that operation, but which more often is experienced afterward, is that of being chased or in some way attacked or frightened by an enemy. Most birds instinctively know what to do when this happens. They run or fly away, yelling for help. But some of the most highly developed birds, such as the crow and jay, have to be taught by their parents to try to escape from an enemy—for instance, a cat. In all cases, whether by instinct or as a learned response, the young bird gives the characteristic fear and alarm cry of his species. He clearly expresses this emotion of fear; and all others of his kind who hear it, as well as many other animals and birds, take warning from his cry and are alert to danger. He may also warn of danger by the flashing of some of his feathers as he flies.

Hunting for Food. Almost as soon as a young bird begins to peck at things, he is beginning to hunt for food, but, of course, his real hunting comes when he is old enough and strong enough to take care of himself alone. His hunting language depends on whether his kind of bird hunts alone or in flocks. Most carnivorous birds hunt alone and give a hunting cry to claim a certain territory. Even birds that hunt only insects and grain have a cry that may claim a territory. Other, more social birds call other birds of the flock to come to eat with them. For example, the light "Tsee-tsee!" cries of a swarm of little bush tits moving through the bushes become more excited when some of them find good insect hunting. They are calling other bush tits to the feast.

Mating. The greatest event in the life of a bird is mating. The first time birds mate, they do so almost entirely from instinct. Later, the more intelligent birds may add variations gained from experience. Though the mating patterns of birds are often quite different in particulars, there are certain basic things most of them have in common. The male usually acts as the aggressor and the show-off, having more brilliant plumage than the female. The female acts coy, shy, and retiring, but usually definitely interested.

The male or female must find a place where a nest can be placed in a protected spot. Then, mainly by voice and display of feathers, but

A male ruffed grouse drums his wings to attract a female.

occasionally by physical combat, the male must win a territory for himself and his mate. In the case of a songbird, he flies to a high spot and sings at the top of his lungs to warn all other birds away. Other kinds of male birds do things like drumming (grouse), gobbling (turkeys), screaming (gulls, jays)—but, whatever he does, a male bird usually lays claim to a definite territory. The female comes and looks over both the male and the territory and nest location. The male

As eggs hatch, the newborn robins open their mouths to indicate hunger.

struts, preens, sings, gobbles, or does anything he can that will attract her attention and make her think favorably of him. When she accepts him, she usually touches his bill with hers and then crouches in front of him. Soon the female is busy gathering materials for their nest, sometimes very simple ones (only a few stones scraped away in the case of gulls) and sometimes very elaborate (a delicately woven hanging nest for the bush tits). The spring is thus a period of much activity and noisemaking.

The Eggs. The sixth great event is the laying of the eggs and the setting by the female and the male. When the eggs are laid, the female lets it be known by her excitement and the cries she gives. The male also becomes excited and makes more song or noise than usual.

Taking Care of the Young. The seventh event in the life of a bird is the feeding and protection of its young. Far more birds than mammals pair for life, and both parents usually take an active part in the rearing of the young. The mother and father answer the hungry calls of the young birds by bringing food to their mouths or regurgitating partly digested food from their own gullets. They are constantly at

A lark sparrow expresses alarm.

work bringing this food, and a conversation goes on between mother and father at this time, telling of food brought and possibly where more is to be found. Other birds, such as the grouse and quail, take their children on food-hunting trips at an early date and help them find food.

At all times the parents are on guard against enemies. The alarm cry from the mother usually brings the father flying home to dive with angry cries at an intruder. The terrible claws of a mother horned owl have missed my head by a fraction of an inch when she dived at me for coming too near her nestlings. Other mothers or fathers, like the mother killdeer described in the first chapter, are adept at luring an enemy away from the nesting site by pretending to be wounded.

SEA BIRDS

Probably the most commonly seen of all sea birds is the gull. Gulls are sharp-eyed scavengers of the seacoasts and some inland lakes. As they fly, they watch each other, and when one sees another gull drop down to the water or on the shore to grab something to eat, it flies in that direction to take part in the feast.

The scream of the gull is used to express many emotions. When a flock dives down on garbage thrown from a ship at sea, the scream is one of excitement and desire, each gull hoping it can get the most choice bits. A more angry and alarmed scream is heard when the gull's eggs or young are threatened, and the bird may even rush at a man, striking with wings and bill. Then there is the call of the male bird to its mate and her answer, a long, keening cry.

Fighting gulls spread their wings about halfway and hop toward each other with short shouts of rage. The blows of the beak are quite savage, until the loser, with a scream of rage and frustration, suddenly turns away and flies, leaving the victor to strut about on the rock and impress the nearest female. The loser in the fight often stops to preen and clean himself thoroughly. This is done by many birds when defeated in battle and seems to soothe injured nerves.

I have watched gulls cleverly steal food from other birds by one distracting a bird's attention, while another swoops in and grabs the food. Sometimes the loud scream of a gull makes another bird drop food, which is quickly swooped on by the pirate.

Nesting sea gulls will attack an intruder.

SHORE BIRDS

Killdeers

Of all the shore birds, the most common and probably the most successful is the killdeer, a black-and-white-marked bird found in grasslands, swamps, and shores of both ocean and fresh water. You have already read in the first chapter how the killdeer mother acts when her young are threatened.

I have heard killdeers in the evening in flocks above me in the gloom, darting erratically through the air, sometimes almost down to the ground, then zooming upward, always uttering an excited "Dee-dee-dee-dee-kee-dee!" These birds seem to be taking part in some kind of interesting game—possibly tag or hide-and-seek—in the sky, their joyful voices seeming to shout, "You're it!" "Now *you're* it!"

On the golf courses and in freshly plowed fields, the constant "Dee-dee-dee!" of the killdeer may be heard early in the morning or in the evening, as the birds cry back and forth to each other while hunting insects and worms. It is a social cry and sometimes a call to better food.

In the early spring mating time, the male dives and spirals and swoops recklessly, calling constantly with wild longing. On the ground, before the female, he dances, pirouettes, and bows, and ducks up and down until you'd think he would be dizzy! But she crouches before him, calls back ardently, and watches admiringly.

On the ocean and bay shores, I have watched little flocks of killdeers and other shore birds gather to probe in the sand for sand fleas, or sand crabs and worms, moving in little spurts and then stopping to dig furiously. At my approach, they gave a shrill cry of alarm and flew as a unit down the beach.

Curlews

Some friends and I saw a long-billed curlew once when we were walking over the alkali flats and marshes near the border of Great Salt Lake in Utah. We came upon a large bird with a great curved bill. Its reddish-brown and spotted coloring so perfectly merged with the russet-colored short grass of the plain that we would not have seen it if it had not suddenly moved its bill and head. At our near approach, it suddenly flung itself forward over the ground, running at first, with wings flapping; then it leaped into the air and rushed skyward, uttering piercing, trilling cries that were repeated over and over.

When we came near where it had been, we saw in a little depression in the grass three large, pale-buff eggs with dark brown spots. The cries of the long-billed curlew now became even more piercing, and you could feel the strong emotion of worry and fear in the voice. The bird circled near us on the ground and began to drag its wing, crying with most pitiful tones. At this sound, the male bird appeared from the east and swept low over us, also uttering the high-pitched cry, rocking its wings and fluttering as if about to fall to the ground. Both birds worked in perfect teamwork to draw us away from the nest and eggs, and, when we were safely away from their sacred place, we heard the emotion in the shrill voices turn to relief and reassurance. At last, crying with a triumphant note, they swept away from us and back to their home.

DUCKS, GEESE, AND SWANS

The great migrators travel far through the skyways, following the spring as it creeps its green northward way toward the pole, or going southward in the fall in search of summer again. They follow the law of the flock because by this law they may live. Usually an old female is the leader, and on her wisdom the other birds depend as they fly. If she is wise about hunters and guns, foxes and coyotes and mink, she will lead them safely through all the dangers of their great migration.

High in the sky, the wise old leader watches for a lake where it is safe to spend the night. Suddenly she gives a cry and points her head and body downward. All the birds of the flock have their eyes on her, so they, too, bend downward almost as one bird and make a long glide to the waters below. Perhaps she sees the glint of sun on metal, half-hidden in the weeds. She gives a different cry, a cry of warning, and the flock rises again in the sky, seeking safer water.

The leader, when a safe lake is found, gives the food-gathering cry, and the flock follows her over the water to the lush seeds of the cattail

Male mallards display while courting females.

or the wild grain (if she is a goose), or to good places to dabble on the bottom (if a plant-feeding duck). As they feed, the old leader and probably one or two others are always on the watch for danger. You see them stop every minute or so to stretch their necks about and look everywhere. One sees the red of a fox's fur in the rushes on the lake's side, and again comes a cry of warning, but a different sounding cry that means "Fox!" in its tone. This time the flock does not fly but paddles out into deeper water where the fox cannot go.

Each species has different ways of mating, but almost always the males try to display their colors and attract the females. I have watched a male mallard duck preen itself before the female, turning to and fro in the water in front of her so the sun will catch the bright green of his wing feathers and flash and dazzle her. It seems as if she turns from him, but usually he herds her back where she must watch him, sometimes flapping his wings and craning his neck far above the water to show her how beautiful he is. All the time he quacks softly to her. When she begins to answer in like tone, then he knows he has made a conquest.

When the young birds break out of the eggshells, they follow the adults like shadows, copying all they do but instinctively scattering to hide in the rushes when the alarm cry is given.

HERONS AND BITTERNS

These birds are solitary wanderers most of the year except during the nesting season, when many species gather in colonies for mutual gossip and protection. Of this group, the great blue heron is probably most likely to be seen. In the spring rookeries where I have watched them nesting in the tops of large oak trees, I could hear the constant "Kronk, kronk!" of gossip between the adult birds, and the shriller piping of the nestlings calling for food. The great stick nests in the tops of the trees were carefully guarded by the parents, who brought regurgitated food to stick down the craning throats and wide-open bills of the young.

In the mornings, the parents usually flew away on a hunt to distant marshes where they would stand like sentinels in the water, their powerful bills cocked and ready to strike. Any fish, crayfish, or other water animal who passed near was swiftly grabbed in the sharp bill and sometimes tossed in the air and then swallowed. The "Kronk!" cry is sometimes given on a note of triumph when the heron is lucky at catching things, sometimes with a deep note of alarm when the bird sees a man approaching. Then it rises clumsily into the air on its great wings.

At mating time, most of the heron and bittern males develop special feathers, more beautiful than the rest, on their necks and heads. These are spread out proudly and waved to and fro in front of the female to charm her. The males sometimes do a kind of awkward dance as they show off these feathers, and the females curve their necks, turn their heads, and look coy.

LOONS AND GREBES

The incredibly fast diving of loons and grebes is part of their language, for they watch each other in the lakes and rivers, and, when one dives suddenly, the others dive, too, because they sense the danger signal. Both grebes and loons are noted for their often fantastic mating tactics.

The Holboell's grebe begins his mating song with quickly repeated, plaintive, and vibrating short wails: "Ah-ooo, ah-ooo, ah-ooo—ah-ah-ah-ah!" Sometimes the song ends with a machine-gunlike chattering trill, like "Whaa, whaaa, whaaa, whaaa, chitter-r-rr-r-r!" It also produces a hoarse and prolonged nasal "Gronck!"—something like the bray of a donkey—and a raucous "Cawack, wawwaacck!" that is very loud. The male, while giving these sounds and others in an amazing mixture, often whirls around in the water at high speed and shows off his head plumes to the female. If attracted, she calls back and swims out to meet him.

HUMMINGBIRDS

Around their nesting places, hummingbirds are often savage tyrants toward any bird or animal who comes too near, diving directly at the object of their wrath and using their sharp bills with good effect. At the same time, they give shrill, angry, and buzzing war cries.

Although most hummingbirds do not sing, an exception is the male Anna's hummingbird, who lives near us and often sits high on a bush

A male calliope hummingbird courts a female.

to send forth a series of thin and cheerful squeaks. When a male Anna's hummingbird goes courting, we see concentrated emotion in action. Down from a high perch he comes, zooming like a bullet, making a vast U in the empty air and giving an explosive whistle with his wings at the bottom of the U. Back and forth and up and down on the U he swings like a pendulum, flashing his brilliant crimson neck feathers in the sunlight. The female quietly sits on a nearby branch and watches him. If she flies off, he goes after her like a shot, and the two gyrate high in the sky in a kind of mating dance.

The male hummers of different species have different shapes to their zooms. For example, the Allen's hummingbird makes a great V when he dives. But in all cases this zooming dive is both a mating song and a warning to all other males to stay away from this territory. Two males often dart at each other with shrill, angry squeaks, striking with their sharp bills. Often they clinch in midair and then flutter to the ground, still fighting, but soon one of them has had enough and flies away to other territory.

HAWKS, EAGLES, AND VULTURES

The languages of hawks and eagles depend somewhat on their methods of catching food. The vulture has claws too weak for grasping prey, so he must soar in the sky, constantly watching for dead animals. He closely watches all other vultures, and his sharp eyes miss no tiny movement in the sky. When he sees another vulture spiral earthward, he flies rapidly in that direction to see what food has been found.

Soaring Hawks

The soaring hawks or buteos (including the eagles) soar high over the land, looking for live animals they can attack. When a rabbit or ground squirrel is seen, down comes the red-tailed hawk from the sky like a thunderbolt to strike its terrible claws into the animal's back. The screams of such a hawk are warnings to similar hawks to stay away from the territory where the redtail is flying and leave his game alone.

Red-shouldered hawks dive at a mouse.

Wild Birds

Harrying Hawks

The harrying hawks, such as the marsh hawks and the kites, sweep low to the ground, particularly over swampy country. They dodge back and forth as they fly, harrying and worrying animals such as rabbits and mice, until they get a chance to grab them. Two such hawks, working together, seem to signal each other how to duck or dive by their wing movements. Their screams are probably territory-claiming cries.

Bird Hawks

The accipiter, or bird hawks—such as the sharp-shinned, the Cooper, and the goshawk—hide in the trees or bushes and suddenly dive out on birds who are feeding nearby. The scream of one of these hawks is intended to strike sudden terror into their prey.

Falcons

The falcons are the great speedsters of the sky. They dive at over a hundred miles an hour on individual birds or flocks. The language of the falcon's movements is often meant to split birds away from a flock by frightening them with bullet-whistling dives. Then the prey can be struck down one by one. Like most hawks, falcons give screams to claim territory near their nests.

Some of the most interesting examples of communication among falcons occur during courtship. To birders the female is called the falcon, while the male is the tercel. At the beginning of the courtship, the tercel tries to interest the falcon in a nesting spot on the cliffs or in a tree. He flies around the cliffs of the mountaintop crying "Wishew" to the female or making creaking noises. Earlier he has been showing off to the falcon by doing high dives at great speed, but when he feels she is getting ready to find a nest and to mate with him, he goes into the most extreme type of flying probably ever seen in a bird. He especially likes to have a high wind to fly against and then, flying high above the mountaintop and cliffs, comes down with the wind in a long swoop at such a terrific speed that you hear a sound like ripping canvas, like an explosion in the wind. Again he flies high, but the next time, when he swoops down the cliff face, he turns over and over in the air, hurtling part of the time or flying completely upside down. Eventually the female becomes interested in finding a nest and begins to go around with him to different sites, but she almost invariably chooses one he has not chosen, a place where there is enough of a hollow for the nest sticks to be held against the cliff. As the first eggs are laid, her interest in the showing off of the tercel wanes, and she becomes more interested in her nest.

Hovering Hawks

The hovering hawks include the sparrow hawk or American kestrel (who is also a falcon) and the white-tailed kite (black-shouldered kite). These small hawks hover over fields on swiftly beating wings, close enough to the ground so their very keen eyes can detect the tiny movements of grasshoppers, mice, lizards, and small snakes among the grass stems. Then down swoops the hawk to strike its prey. Often one member of a pair screams to the other to come join it in the feast. The sparrow hawk, or kestrel, has a shrill whistle, as does the kite, that may be either a warning of alarm or a challenge.

Many times a hawk has screamed at me when I approached its nest. The sound, full of alarm and worry and challenge, calls its mate. Then the two of them circle about screaming repeatedly until I go away.

The mating language of most hawks consists mainly of motion, though occasionally two scream in unison. Usually the male marsh hawk, for example, sweeps in great circles high above the female, then plummets earthward, somersaulting several times; just missing the ground, he swoops upward again. The rushing sound of his wings is like a whistle in the air. Soon the female joins him. In the case of some hawks, such as the duck hawk, the two may fly clear out of sight up into the blue and then come tumbling down together.

NIGHTHAWKS AND SWIFTS

The nighthawks—including also the poorwills and whippoorwills—are creatures of the gloom, seekers of moths and other night-flying insects, whom their wide-spread and well-haired mouths are especially adapted to catch. The swifts are dazzling fliers of the daytime.

I have heard the poorwills calling "Poo-woo, poo-woo" in the early dawn, a cry of mate to mate and of territory claiming. In the dense brush of the California chaparral, I once came upon a poorwill suddenly, so that it rose up from under my feet like a great, fluttering moth. The whippoorwill of the East and Southeast is also a brush- and tree-dwelling bird, and its soft, lovely, and plaintive cry rises in the summer evenings from along every roadway.

In the great plains of Colorado, I have watched the nighthawks erratically trace pathways through the evening sky, then suddenly zoom to earth and boom their wings like a thunderclap, all of them doing this together as a sort of pattern or game, even while they were catching insects. While they play in this way, they call to each other with a kind of nasal, unmusical squeak, a social cry of friendly companions, like the shouts of boys and girls playing tag.

Over the Great Plains by day, I have watched the swifts perform whirling, sweeping dances between me and the sun. Time and again dozens of the birds swept across the sky in great half-circles at dazzling speed, over and above each other; they then tumbled downward through space, only to swoop up again straight into the sky so fast that they changed in a few seconds to almost invisible specks. All the time their excited and spirited chippering came down from the heights. This, in the language of the swifts, means fun, joy, companionship, a game they are playing even while catching high-flying insects on the wing with incredible expertise.

DOVES, PIGEONS, AND QUAIL

Wild pigeons and doves have developed signals of alarm and motions of defense against enemies that tame pigeons rarely need. The white tail feathers of mourning doves flash in the light like mirrors when a pair of these birds take to the wing in startled flight. The flashing feathers, along with a sharp cry of alarm, send a signal to nearby doves that danger is near. As they fly, these doves also give out from their wings a clear, whistling sound that may sometimes serve as a warning, sometimes a social gathering signal.

The "Coo-coo-coo" of the mourning dove male is a delightful song of late spring and early summer. The male also seeks to impress the female by suddenly rising with violent wing flappings straight up into the sky for a hundred feet or so, then gliding back on nearly motionless wings. The two mates often express their connubial relations by long periods of caressing each other's bills. It has not yet been scientifically established whether gestures of this kind are displays of affection or simply instinctive.

While flocks of pigeons and doves seem to be held together by little more than love of companionship, the quail and bobwhites are much more cooperative in relation to danger and food. "Wheee-oooo!" whistles the California quail who is on guard when an enemy appears, and into the bushes the flock runs, or takes off on whirring wings for distant parts.

Quail and bobwhites are both great conversationalists in their flock movements. Quail scold each other with a "Ka-ka-ho!" cry. Soft "Cut-cuh-cuh!" notes of companionship keep the flock together. In querulous notes the young birds ask their elders where the food is. There is the loud gathering cry of the clan, the "Bobwhite!" and the "Quer-ka-go!" that bring the flock together after it has been scattered by a fox or dog. At night the flock forms a circle with heads out, all ready to fly, while a little humming, happy note, a kind of lullaby, sings them to sleep, except for one sentinel on watch.

In the spring, the flocks are broken up as mating takes place, with the usual fights between the males over territory and females. Later, the father helps stand guard over his family. At his sharp, whispering cry of warning, the young scatter like little leaves in the grass and are hard to see.

OWLS AND ROADRUNNERS

The great horned owl has been aptly compared to a ghost because of its ability to fly softly through the air; no sound is heard until its claws close on a luckless rat or rabbit. It lives mainly in the woods, and, like most owls, each individual has a definite range that it calls its own. The deep "Hoo, hoo-oo, hoo-hoo!" of one horned owl is a signal to all others that this is its territory and they must stay away.

A short-eared owl chases a rabbit.

The sound, mysterious and full of menace, may also be useful in striking such terror into the hearts of small animals that they run foolishly into the open.

The horned owl squalls like a cat sometimes during the mating season, probably calling to the female while also challenging or threatening all other males. Doglike yelps are also heard at this time and sound like the eager cry of the male seeking the female. Only once have I heard an owl scream in what must have been fear, and that was when a red-tailed hawk attacked a horned owl in the top of a tall eucalyptus tree and knocked it to the ground. But its blazing yellow eyes and snapping beak warned me not to come too close.

The scarecrowlike roadrunner is one of the fastest of all running birds and an expert at catching snakes and lizards in the desert country. Roadrunners have been so hunted in most places that they are very wary of men, but in wild country I have seen them show considerable curiosity, cocking the head from side to side and eyeing me curiously.

One morning, just when the sun rose in the desert, I awoke in my sleeping bag to hear an extraordinary uproar of hoots, chirps, cooings, and many other odd noises almost impossible to describe. A roadrunner hopped out of a big bush where the noise had come from and ran off. It was certainly performing some kind of song to woo a female and probably to claim territory for itself. But this song is rarely heard.

WOODPECKERS AND FLICKERS

The merry hammering of woodpeckers and flickers, all the way from the light "Rat-a-tat-tat!" of the tiny downy woodpecker to the loud thunder of the big pileated woodpecker, forms a language the American wilderness would sadly miss. It is the talk of sturdy, independent birds who work hard for their living by cutting holes in bark and wood to get at the insects inside. The beating sound of the bills is a message telling other woodpeckers, "This is my tree!" The bright colors on the heads and bodies of most woodpeckers are also message senders, telling what kind of woodpecker is working and warning other woodpeckers to stay away. Of course, in the mating season, the bright colors are used to attract the females.

A hairy woodpecker sits outside a nest hole.

In summer most woodpeckers are rather quiet except for their hammering. But in the springtime the trees are often filled with their chattering cries as well, while their wings flash bars of light as they fly constantly back and forth. The males are courting the females, and their chattering means either that they spy a prospective mate, or that they are exasperated and angry because they see a rival male. Many times you will see one woodpecker chasing another round and round a tree in spiral fashion, moving with jerky motions. After weeks of this excitement, chasing, and some fighting, the mates are chosen and the territories decided upon.

PERCHING BIRDS

By far the greatest number of birds belong to the order of perching birds, which includes all the sweet singers. These birds have a highly developed sense of territory, and the song of the male of each species tells all other males that he owns the neighborhood. Some birds sing a different song later that seems to be a song of joy in nesting.

The other sounds these birds make are mainly social calls of mate to mate or to members of a flock. They may be war cries between fighting males or cries of alarm when an enemy comes near. It is often possible to decide what the sound means by watching what the bird is looking at when it cries.

Flycatchers

Flycatchers, or kingbirds, are without the territorial song of the songbirds, but they do indicate territory with call notes and occasional chattering war cries. Kingbirds even sometimes seem to scream with rage when they fly at a hawk or crow who comes near their nest.

A flycatcher always signals that it is a flycatcher by the distinctive way in which it catches insects. Most flycatchers sit on the outer ends of tree limbs watching for flying insects. If one is seen, the flycatcher suddenly springs into the air, snaps up the fly in its wide beak, and circles back to the same branch from which it sprang.

Though not necessarily very intelligent birds, the kingbirds are among the most interesting because of their indomitable will. When they make a nest and lay eggs, they become guardians of a sleeping place that is quite exposed to view. They become fierce warriors when that nest is threatened. They ignore most birds and small creatures as beneath their dignity to attack; even some humans are placed in this class, those that mind their own business and show no sign of approaching too close. But let a man come close with a camera or a gun and he is at once marked as an enemy! The scream of the tyrant flycatcher male rises to a screech as fierce as the roar of a tiger, announcing his attack on the one who dares to come near. If a small enough bird or animal comes close, it could even be killed. A man may feel his head struck by bill or claw hard enough to draw blood!

Wild Birds

A crow is harried unmercifully until, dodging desperately, and cawing for help, it gets away from its enemy. A dangerous bird hawk, however, is repelled more carefully: the scream of the kingbird is just as loud as it dives at the hawk, but it stops just short of striking the bigger bird, because it knows a hawk can turn swiftly and present its dangerous claws. This conversation between the kingbird and all possible adversaries says quite loudly, "Get away from my nest quick, or be hurt!"

Young kingbirds, who call when hungry, are looked after carefully by the adults and well fed with worms and other small creatures, but as soon as the young take on adult plumage, they are driven away from the nest to take up life alone. They are usually well-equipped to meet the world and soon ready to take up their own war cries, perhaps even when they fly south in the great bird migration of fall to warmer Mexico and Central America.

Crows and Jays

Crows and jays are among the most intelligent of all North American birds. They certainly have what seems to be the most extensive language, of which only the highlights can be given here. These birds need to teach their young the tricks of life and talk much more than most other birds, which means also that they can learn new things more easily. A young crow does not instinctively know the appearance of an enemy the way a young robin does. The crow youngster has to be taught fear of a cat by the way its parent reacts to such an enemy.

On our ranch, the flocks of crows pass back and forth almost daily on their way to and from their roosts and feeding grounds. The clear "Caw, caw" of the crows sounds from the sky and is used as a signal to tell of food found and to draw the flock together. A harsher, shriller cry of alarm may suddenly speed up these big black birds, and, if they are shot at by guns, they tend to scatter and intelligently put as many trees as possible between themselves and the shooter.

In attempting to spy into the nests of crows on cliffs, I have been called a wide variety of uncomplimentary names by the parents and even suddenly attacked by them. It was amazing how the inflection of the "Caw, caw!" cry could be changed and charged with different

Crows mob a saw-whet owl.

emotions of hate, anxiety, fear, and despair as I approached nearer and nearer to the precious nest.

Crows become very loud and noisy when they find an animal they feel they can safely bully and pester. I once saw them mob an owl and peck the poor creature to death, their screams sounding like those of an emotional human mob at a lynching. Crows have also been known to attack one of their own members and kill it, after a noisy conference that has reminded some observers of a courtroom scene, with a judge and jury. Whether a crow law had been broken or whether the flock had decided to get rid of an old and sick crow could not be determined.

When crows feel they are unobserved by man, they carry out many seemingly complex conversations and also have been heard to warble a little song that may be an expression of affection between mates. The movements of crows have rich meanings to other crows. In the flock there is a definite peck order, very similar to that found among chickens. The best fighter is at the top of the social order; he gets the pick of the food and first choice of a mate. When two crows fight, they caw angrily, bristle their feathers to look bigger, wiggle their wings, and strike with beak and claw, often tumbling over and over. The winner can thereafter always drive the loser away by just bristling. Even the expression of the eyes is important and shows love or hate, trickery or playfulness.

The king crow seems to lead the flock, but sometimes there is angry disagreement about where to go or what to eat. In the face of danger, however, such quarrels stop quickly. A crow flock attacked by a duck hawk will suddenly bunch so closely together that the duck hawk does not dare dive into the mass.

I am convinced that some birds, including the jays, titmice, crows, blackbirds, and probably several other kinds, have a certain amount of real intelligence and even foresight. One Steller jay in the mountains learned to attack a cat just as it was coming out through a swinging screen door at the moment it was temporarily helpless to fight back. The jay seemed to anticipate just when the cat would come, would fly down to strike at it with wings or claw, and then fly away with a loud and rapid "Tchay-tchay-tchay!" of triumph. Actually, individual birds vary greatly in intelligence and ability, and each within a species has a distinct personality, so that the variations of language used are far more rich in detail than could ever be told in this book.

All jays are alarm callers when a new animal enters their forest or hillside, and small animals take warning from these notes and run for their lives. By the tone of the sound, you can tell whether something large and dangerous is coming or just an animal they think they can pester. When jays themselves turn into nest robbers, they become suddenly very silent and sneaky in all their actions. However, when robbing squirrels of their hidden nuts, jays shout with triumph and call names at the squirrels as loudly as the angry squirrels shout at them!

I have heard jays squeal like rabbits, and even make a noise like a car motor; but I became most excited when I heard a crested jay singing. It was a weak, warbling song, but nevertheless a tuneful one, probably sending a message to a mate.

Other Perching Birds

Most of the other perching birds are real songsters, using their songs not only to show the territory each male feels he owns, or to lure or charm a female, but also just bursting forth with the joy of living. In the springtime, there is much spreading of wings and feathers and other displays of males at the edges of their territories. If these displays do not drive the rival away, there may be a fight or half-fight, with two birds clinching and pecking furiously. You often see two birds flying in and out of the bushes, first one chasing, then the other. Many birds, like the towhees and wrens, make a buzzing cry of anger at such times.

Most of these birds mate for life, but between the mates all is not necessarily smooth sailing. Birds who become very angry at each other often make so much noise about their nesting and raising of young that they attract enemies who eat the young. Wiser couples learn to be quieter. Males urge females to build nests but rarely help. Once a male titmouse who had entered an old woodpecker's hole in a tree urged his mate to use this as a nesting spot by making noises like a baby bird, but she turned him down, much to his annoyance, and he scolded her sharply. The female, in her turn, became much annoyed at the male later when he did not bring her food as often as she wanted when she was sitting on the eggs. At this time, she called her mate with noises like those of an older nestling bird. This use of baby talk by both male and female at nesting time is common among many birds, and is useful in conveying ideas. The fluttering of wings close to the body and the making of baby talk is often used by a female to attract her mate, while the male sings and displays his wings and tail feathers to attract the female.

Many perching birds signal alarm to each other as they fly by the flashing of white tail feathers and other kinds of markings.

Tiny movements may have meaning to birds that are missed entirely by a watching man, and much careful observation and study of bird language remains to be done.

7
Reptiles, Amphibians, and Fish

Because their brains are smaller and because most reptiles, amphibians, and fish let their young shift for themselves, almost as soon as born, the languages of these animals are much less rich than those of mammals or birds. Also, what language they have is almost entirely instinctive—that is, without conscious thought behind it. However, there is a tremendous amount yet to be learned about these languages, and even what is known would probably fill a book.

REPTILES

Turtles

Turtles seem to range in intelligence from the rather stupid but extremely dangerous and powerful snapping turtles to the comparatively intelligent and gentle tortoises. Tortoises make excellent pets and even show affection toward a master. They communicate this affection by trying to crawl onto his foot, or even by resting a head against his leg and looking up into his eyes.

The alligator snapping turtle has a tonguelike device in its mouth that looks like a white grub; the turtle wiggles this as it lies still on the bottom of a river or pond; it thus lures a fish close enough for the turtle to snap out its neck and jaws like a striking rattlesnake. The snapping turtles and soft-shelled turtles are noted for their viciousness. They warn that they are about to strike by pulling back their head and neck and gazing balefully with their eyes.

Very few turtles make any sounds beyond a slight hiss of warning, but some of the tortoises, who live on land, are known to make repeated rasping noises over and over in the mating season, probably calling the females. Some male aquatic terrapins—turtles who spend all their lives in water—show a remarkable courting behavior. The male swims beside the female underwater, and, as he does so, reaches out with one of his front arms and, with a long, peculiar-looking nail on one finger, tickles or strokes the female along her neck.

Most turtles have learned from sad experience to keep a constant lookout for enemies and to pay attention to the slightest warning movements of their own kind. Many a time I have sneaked up to a Pacific pond turtle only to have it suddenly plunge into the water, with others nearby instantly following. Most turtles pull their legs and head inside the protection of their shell when attacked by an enemy, but only the box turtles have complete protection. Their plastron, or belly cover, positions several horny plates in a continuous shield that is held in place by powerful muscles.

Musk turtles have glands that give out a powerful odor, a bad smell that helps discourage animals from eating them, and which is also a signal of the animal's fear and anger. It is also used to let other musk turtles know of the presence of one of their kind, and as a calling card to be sent to the female by the male.

A female turtle found digging in the sand or ground of a beach or bank is almost certainly preparing to lay eggs. She covers them carefully and then leaves them to incubate.

Alligators

Alligators are found wherever there are warm and humid swamps in the southern states. They are very much afraid of men and disappear quietly and quickly into the water when they hear a man approach. The big bull alligators can produce the loudest noise of any reptile: a deep, booming bellow. This sound echoes from the swamps most loudly at mating time. Besides his bellow, the bull alligator sends out wave after wave of fetid alligator smell from his scent glands, presumably to attract his mate. A grunt of warning is another alligator sound that is occasionally heard. It will send a group of basking alligators plunging into the dark water. Alligators have attacked swimmers.

Reptiles, Amphibians, and Fish

Lizards

A lizard who is watching for an enemy will usually move its head from side to side in short jerks while its body is lifted off the ground. If it sees something it considers dangerous, it scuttles away at full speed or even lifts its body higher and runs on its hind legs (as collared lizards and some others do). The instant other lizards notice it is running, they too run, though they soon stop and look around alertly to see what caused the disturbance.

Certain lizards of the iguana family, particularly the common fence lizards or swifts, signal each other with the colored markings on their bellies and throats. Most fence lizards have blue bellies, which, in the males, are deeper in color and extend up to the throats. Each male fence lizard claims a territory by standing on the highest point of a pile of rocks or sticks and pushing its body up and down with its front legs. This flashes the blue belly in the sunlight and acts as a

Anoles and fence lizards.

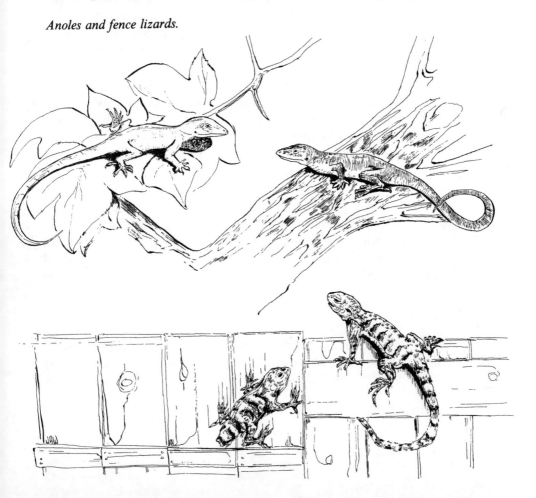

warning to other males to stay away, while the signal at the same time calls to any female who is near to come to investigate.

Most lizards, when cornered, open their mouths wide and give a hiss. Alligator lizards, though seldom much more than a foot long, hiss quite fiercely and open their mouths so wide, to express their warning and anger, that they no doubt scare many animals that have started to attack them. I know a puppy who once jumped at one of these lizards, and was so impressed with its formidable appearance that he stopped his attack.

Many lizards also communicate by means of smells. Fence lizards have glands along their thighs that they rub on the sticks and rocks over which they pass. These scent messages undoubtedly tell other lizards the nature and sex of the lizard who left them.

The bright colors of the gecko lizards of the Southwest are useful in imitating brightly colored poisonous creatures like the Gila monster and the coral snake, and so protecting themselves from attack. The American chameleon or anolis seems to show inner feelings and moods by the color of its skin. When it feels peaceful and happy, its skin is usually brown, but if it becomes angry or frightened, it often turns bright green. These lizards also display brightly colored throat fans to attract females and to warn away other males.

Most lizards are insect eaters. You can often tell when they are hunting insects by their catlike creeping over the ground. When near a fly, the lizard gathers his leg muscles together, quivers slightly, and then rushes forward to grab his prey.

Most lizards are able, when pursued, to leave their tails in the clutches of their pursuer. And the alligator lizards are often able actually to throw their tails loose when they wish to distract an animal who is chasing them. Horned lizards rarely lose their tails, but some of these remarkable animals can suddenly increase the size of the eyeball, make the neck rigid, and then, giving a rasping sound, shoot a stream of blood from one or both eyes. This should indeed startle an enemy!

Snakes

A good deal of the language of snakes seems to be carried on by means of smell and is therefore largely untranslatable to men. Most snakes leave their smell along the ground as they travel over it, particularly in their droppings, and this smell is detected by the forked tongue of another snake. As you watch a snake move, you usually see this black, forked tongue stick out at intervals to test the ground. A male snake can easily follow a female who has left a smell trail, and he will know from the smell whether she is ready for mating and, if she is, will show his increased excitement by the rapid flicking in and out of his tongue. Most snakes also track their prey down by smell, striking at it when they get near enough to see it move.

Most snakes signal their intention to strike by drawing back the head and partly coiling the body. The rattlesnake, of course, uses its rattles to warn potential enemies to stay away but keeps the rattles quiet when it prepares to strike its living prey.

The comparatively high intelligence of the very successful garter snakes was illustrated by what happened when a group of them was thrown a frog for food. The snake who caught the food wriggled its tail at a great rate, making so much noise and disturbance that the other snakes were distracted and didn't notice that it was eating the frog until too late! Garter snakes also make considerable use of a bad smell that makes most animals glad to leave them alone.

Many comparatively harmless snakes assume very threatening attitudes when they are attacked, but these are mainly bluff. The common water snake is so sinister-looking, especially when it flattens its body and hisses viciously, that many people take it for a poisonous snake. Another snake that fools many into believing it is dangerous is the common hognose snake (also called puff adder) of the central United States. This snake startled me one day in Colorado by puffing up its body to about five inches thick and hissing at me venomously. But when touched, it rolled over on its back and pretended to be dead, opening its mouth in a ludicrous way. When turned over on its belly again, it immediately flipped back onto its back, as if insisting that this was the only proper way for a dead snake to look.

Certain snakes that superficially resemble rattlesnakes, such as the gopher snake, often beat their tail rapidly in dry leaves to frighten an attacker. The gopher snake or bull snake has a ferocious loud hiss that it uses very effectively for bluffing purposes. Another snake that is a successful bluffer is the western ring-necked snake, which twists its tail into a corkscrew, displaying the bright red underside, and frightens its captor into thinking it is poisonous.

It is interesting to watch a snake catch a mouse. If it becomes

Snakes: (clockwise from upper left) *western rattlesnake; western hognose snake; western hognose snake playing dead by turning over on its back; black rat snake; and* (center) *ring-necked snake.*

interested, its tongue comes out and flicks around more and more rapidly, as it smells the mouse. Then gradually the snake's head begins to move from side to side, following the movements of the mouse. If the mouse comes near, the snake will suddenly strike it, seize the prey in its jaws, and throw a loop of body over it to hold it down to kill it, either by a bite or by constriction. The snake licks the mouse's dead body all over and then swallows it whole by the wonderful snake method of detaching its lower jaw from the upper.

AMPHIBIANS

Salamanders and Newts

People without knowledge of nature often mistake a salamander for a lizard. The difference between the two animals is that the lizard has a dry and scaly skin, while the salamander has a moist, smooth skin. Salamanders and newts are generally found in moist woods or mead-

Reptiles, Amphibians, and Fish

ows or in streams or ponds, and when the weather becomes very dry, they often hide in holes in the ground or under rocks and bark.

Most salamanders twist their bodies and try to escape when you pick them up, but the slender salamanders of the West Coast wriggle and lash so violently when touched that it is quite difficult to grab them. Many salamanders lose their tails if roughly handled. This helps them escape enemies, particularly the snake, who may grab the tail end only to find it coming off in their mouth while the salamander runs off.

A few salamanders make squeaking noises when disturbed, and the Pacific giant sends out a screaming and a rattling that warn you to leave it alone. Some salamanders, when bothered, arch their backs like bows, with the midsection bowed down and the head and neck up. Apparently this is a bluff to make you think the salamander is poisonous. If touched, the Eschscholtz salamander of California not only bluffs in this way, but rocks back and forth on stiff legs and gives off a milky, sticky fluid that stings if it touches human skin.

Different male salamanders have different methods of courting females. The Eschscholtz salamander, for example, rubs his body over the female and gets her to follow him into water by lashing his tail over her neck. The tiger salamander male and female may nip each other, do much body rubbing underwater, and lash their tails about each other until the water boils. Most salamanders mate and lay eggs in ponds or streams, but the arboreal or tree salamander of the West lays its eggs in damp places in the woods, often in the crotches of trees.

Frogs and Toads

Frogs and toads are much more noisy than salamanders, but most of their noises come from the male when he is attracting the female in the mating season. Thus the peeping of small tree frogs and the deep "Jug-o-rum!" of the giant bullfrogs have the same function. There is undoubtedly a lot of joy of life in the sound, plus a social feeling of being together in a great songfest.

The males usually gather in the mating pools before the females come, and there they begin calling them. After a while the females may answer from the land with lighter voices. When they do, the

A male Pacific tree frog sings to court the female.

males usually greatly increase their excited chorus. When the females enter the pool, there is usually much pushing and shoving among the males, but no actual fighting. The female usually accepts the male that gets to her first.

Frogs and toads make various croaks, squeaks, chirps, and similar noises when frightened or captured. These noises serve to warn other toads and frogs and possibly to bluff or startle whatever has seized them into dropping them. Some toads, in particular the giant Colorado River toad, make clucking noises like chickens when they are sleepy and happy. Bullfrogs are noted for their almost humanlike scream when hurt. They also give loud yelps when startled. Some bullfrogs pretend to be dead when captured but quickly hop away when let go.

FISH

Fish that swim in schools, like trout or shiners, communicate with each other by the flash of their silvery scales. The turn of a frightened fish's body, as the tail beats swiftly, usually causes the silvery under-scales to flash a warning to nearby fish in the school. The whole school then dives downward toward safety. Certain fish may also discharge a smell into the water, which warns all nearby fish that danger has come. They then retreat from the source of the smell.

The courting of the female by the male in many fish is quite elaborate and is very distinctive for each species. The male and female each respond to what the psychologist calls sign stimuli given by the other and go through a careful procedure in great excitement. The stickleback male, for example, starts the breeding season by establishing a territory over which he is ruler. He knows another male is approaching if he sees the red color of its belly. He immediately stands on his nose on the bottom of the stream or pond and waves his fins. Usually this warning is sufficient to send the other fish away. Otherwise the first attacks the other male by hitting him with his snout, particularly if the second fish also stands on his nose on the bottom. After a few bites on body and fins, one of the fish swims away in defeat.

As soon as the male sees a female, he acts very differently. He knows her by her large plain belly, and he shows off his red belly to her, beginning a zigzag dance. This is as if he said to her, "See how handsome I am and how well I can dance!" The female responds by sloping her body down toward him in the water. The male then dances around the female as they swim down into the water, and the female swims into a nest of rushes that the male has built at the bottom of the pond or stream. Here the male nudges her with his nose, and in obedience to his command, she lays her eggs.

As most fishermen know, fish telegraph their intention to feed by their more active motions and their interest in small objects in the water. To catch an insect, they "strike," which means simply that they make a sudden quick rush with wide-open mouth and then snap it shut on the prey. A fish telegraphs its dislike of being caught with a hook by leaping high and shaking its head violently from side to side to get rid of the obstruction. Some wise old fish have learned from harsh experience to approach every new thing with great caution. You can see them eyeing a shiny new lure and then lazily turning away with a contemptuous flip of their fins, as if to say, "You can't fool me with that thing!"

Sometimes, when chased by a mink or otter, an experienced fish will suddenly flash its silvery belly toward its pursuer, then quickly turn in such a way that the silver disappears and the fish merges with a dark background. The mink is lured by the bright color and overshoots its mark. By the time it has turned, the fish has disappeared. Catfish, swimming through muddy river water, keep the barbels or feelers on their heads stiffly erect, seeking to touch animals they can catch and eat.

Fish may communicate by expelling air from their bladders into the water or scraping their gill spines together. These may make sounds that say, "Here is food!" or "Look out, an enemy is coming!" We can discern the meanings by watching their actions. Some evidently "talk" by touching each other's lateral line, a sensitive area on the sides, or by pulsing water at each other.

8
Insects and Their Relatives

The world of the insects, with its hundreds of thousands of different species, presents another field in which little study of animal communication has been done. I hope what is said in this book will encourage the reader to take part in the future exploration of this marvelous miniature world.

Solitary insects, such as grasshoppers, plant bugs, and most of the beetles, have little need for communication except at the time of mating. However, many of them have warning colors and signs that tell us they are dangerous or potentially dangerous. The brilliant, metallic colors of a wasp warn us of its sting. But there are also perfectly harmless insects that copy the harmful ones in order to use their colors to escape from enemies. Several flies and moths do this copying of the more dangerous wasps and bees, and are hard to tell from them until you look closely and see the two wings of the flies and the hairy scales of the moths.

Many insects communicate their emotions with the sound of their wings. Flies buzz with a particularly loud, angry note when they are caught in a spider web. They seem to give a lighter, buzzing cry of joy when they break free from the web and fly away. A still different and peculiarly intense buzz is given at mating time.

The loud buzzing of male cicadas on a summer day is a call to mates, as is also the trilling of tree and common crickets. The noise is usually made by rubbing a leg against a wing at high speed. Only insects that use noises to attract one another in this way have genuine ears. The ears of a cricket are in its legs!

Other insects talk to each other at mating time by smell. Many

Male crickets singing.

male moths have featherlike antennae, which they use for detecting the smell of the female sometimes at a distance of a mile or more. Smell is also used by certain insects for protection and warning. The various kinds of stink bugs give off very unpleasant odors when handled, while the dark-colored bombardier beetle lifts up his rear end and shoots a jet of ill-smelling gas.

Some insects, such as the wasp and bee, buzz, while others, such as the velvet ant (a wasp, not an ant), and certain of the true bugs, squeak their warnings. The squeak is made by rubbing the legs. Many grasshoppers, as they fly, make a loud clicking noise that both warns other grasshoppers and startles enemies.

At mating time, many male insects try to impress the females with strange movements, dances, and bright colors. Colors may also be used to confuse an enemy, as when the bark butterflies suddenly fold their wings and look so much like part of the bark that they cannot be seen.

BEES

Social insects naturally need more of a language than solitary insects in order to run the affairs of their communities. But many of these languages are still completely unknown or very incompletely known. Here, for the purpose of example, it seems best to discuss one language that is fairly well known, the language of the bees. Someday naturalist explorers will make the languages of ants, termites, and social wasps equally well known.

The language of bees is quite complicated, but this does not necessarily mean that bees use reason and intelligence in communicating with each other. Instead, as with most animals, their language seems to be largely instinctive and emotional.

The average beehive numbers about fifty thousand to seventy-five thousand individuals, equal to the population of a small human city. In the hive, during the warm months of the year, the bees are intensely active, bringing in food and storing it, and feeding and taking care of the queen, the young bee grubs, and the helpless drones. There is also the activity of building the honeycomb, keeping the beehive clean and ventilated, and repelling any enemies who might enter. All this requires various means of communication.

Bees talk to their grubs through the language of food. If they want a grub to grow to be an ordinary worker, they feed it bee milk—a mixture of honey and bee saliva—for the first two days, but then give it a pap of honey and pollen. If they want the grub to grow up to be a queen, they move it to a large cell and feed it only bee milk, which makes it grow to several times the size of the ordinary worker.

Bees have bee nurses who look out for the young grubs that live in the cells of the honeycomb. When these grubs get hungry, they begin to wriggle, and this is the signal for the bee nurse to bring the babies their honey and pollen food. When the grub begins to form itself into a motionless pupa before it turns into a real bee, this is a signal for the bee nurse to come and seal up the entrance of the cell with beeswax.

Other bees are living fans who stand near the entranceway and vibrate their wings so fast that it creates a current of air all through the nest to keep it ventilated on hot days. New bees join to help with this ventilation job if the day gets hotter. Still other bees are guards who stand at the entranceway and check on everybody who comes in and any animal that comes too near the hive. They can tell by smell if any bee or other insect who comes to the hive entrance is a stranger and an enemy. Instantly they buzz with anger and rush forward to sting the newcomer.

When a hive prepares to swarm, the workers seem to know they must produce a new queen. They hang wax cups near the center of the brood-rearing area and get the old queen to lay an egg in each cup. These eggs soon hatch into little white grubs that are fed only bee milk by the workers, and this turns them all into queens. The old queen is not fed so much food as before, so that she grows thin enough to be able to fly. After the first new queen emerges from her pupa and her cell, the bees all become very excited and rush around, apparently deciding who is going to swarm and who is not. Those who are going to swarm gorge themselves on the honey stores, and then thousands of them rise in the air with an excited buzzing of myriad wings. They take the old queen with them and fly away with her to find a new home.

The new queen, as soon as she becomes mated with a drone, or male, rushes around and stings to death all the other young queens, unless some of the workers prevent her. If they do this, she may form her own swarm of bees and fly away with them, leaving a still younger queen to keep the old hive going. Generally new queens fly up into the sky, and the drones follow them, so that they mate in the sky and then come back to the hive.

The most amazing and complex part of the bees' language is when

Honeybees: (left) *worker bees indicate the direction of nectar by a "round dance";* (right) *a bee gathers nectar;* (below) *bees work on a honeycomb.*

a returning bee tells the others in the hive where to look for food. Suppose the explorer or scout bee has found some good honey flowers about sixty yards from the hive. Almost as soon as she enters the hive, she begins to regurgitate some of the nectar. The taste and smell immediately tell the other bees the type of flower she has found. They also detect its smell on her body hairs. If she acts excited about it, they know she has found a good place to get nectar. She shows both the distance to the food and her excitement by a "round dance," circling first to the right and then to the left, repeating this vigorously. This dance tells the other bees, "There is plenty of food near the hive!" They begin to fly out in all directions but look only near the hive until they find the nectar in the flowers.

If the flowers found by the bee are farther away than about one hundred yards—say two hundred and eighty yards from the hive—she does a different kind of dance. In this dance she wriggles her tail violently and makes a complete circle to the right, moving forward, then in the original direction. Next she makes a complete circle to the left and moves forward again in the original direction, still wriggling her tail. The direction she moves in a straight line, in this "wriggling dance," shows the direction for the other bees to fly to reach the nectar, since this movement is always made in relation to the direction of the sun. If the food is found in the same direction away from the hive as the sun is, the bee makes her wriggling dance go straight up the vertical hive. If the food is found in the opposite direction from the sun, then she moves straight down. If, during the straight part of the dance, the bee moves, say, fifty degrees to the left of vertical, the nectar she is telling about can be found fifty degrees to the left of the sun, and so forth.

In this remarkable dance language, the bee not only tells the direction of the food she also tells the distance. Experiments have shown that she tells the other bees the distance of the food by how fast she makes her circles in the wriggling dance. If it is not much farther than a hundred yards to the good flowers, she does the wriggling dance circles very rapidly, often as many as nine circles in a fifteen-second period. If farther away, she does the dance slowly; for example, she makes only two circles in fifteen seconds if the food is about a mile away.

The other bees watch and smell the scout bee to get her message;

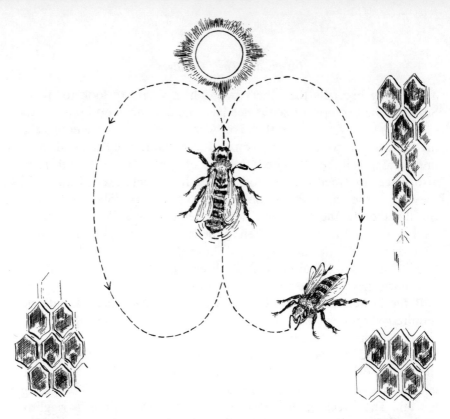

In the bee's "round dance," the speed of the circling and the angle of the straight line in relation to the sun indicate the distance and direction of the honey-bearing flowers from the hive.

then they leave the hive in a steady stream. All but the youngest and most inexperienced bees fly in the right direction and go the right distance to reach the discovered flowers! Is this not a remarkable language?

Bees do not hear very well through air, it has been found, but they have sense organs on their legs that help them hear through solids. Thus, in a hive, they hear sounds brought to them through the walls of the honeycomb. If they hear a loud buzzing noise at the entranceway, they know an enemy has come there, and they may swarm out to fight. The queen occasionally makes a strange piping noise, which may mean that she is hungry or perhaps that she is ready to swarm. Also, two queens may make this noise when they start to fight each other.

The smell of the queen is very important to the bees in the hive. This smell spreads to all the bees and gives them their own distinctive

Honeybees chase a moth from a beehive in a hollow tree.

hive smell that helps them distinguish friends from enemies. The smell of the queen tells them when she needs more food and when she should be prepared for swarming.

SPHECID AND OTHER SOLITARY WASPS

The solitary wasps, which often look like the colonial wasps but do not build large nests in which many wasps live together, are not usually considered as intelligent as the social wasps. Many of them make cells out of mud, which they fashion into little rooms where they lay their eggs and then hunt for prey such as spiders, cicadas, grasshoppers, and plant bugs. Since most of their prey are harmful insects, the sphecid wasps are generally considered useful to mankind.

One of the problems many sphecid and other solitary wasps have are parasites. The parasite, usually a fly but sometimes a parasitic wasp, watches where the sphecid wasp is placing its mud cell for a nest. When the sphecid wasp leaves its cell, with its eggs or some stung or paralyzed insect prey in it, the parasitic insect quickly flies in. It lays its own eggs in the cell, which grow up and eat both the wasp eggs or larvae and the insect prey that was placed there for the sphecid larvae. The parasite is saying, "I am stealing your cell for my larvae, who will feed on what you have placed there!" However, some sphecid wasps are clever. They understand what the parasitic fly or wasp is doing and chase it out of the cell before it can lay an egg! They are saying, "This is my cell and my young eggs and larvae. Get out!" The most clever of all the sphecid wasps do not merely chase the parasites out of their mud cells; they place fake cells here and there and pretend to lay eggs in them. This behavior says, "I'll fool you good!" and it does!

SOCIAL WASPS

The social paper-making wasps are usually large, conspicuous, lively colored wasps. The genus *Vespa* includes hornets and yellow jackets. Hornets are large black and white, yellow and black, or bald faced insects, and are among the most dangerous of all our insects, with very hot stings and nasty tempers when disturbed. Their thickness and large size plus their fierceness makes them difficult to watch. They usually produce large, round- or oval-shaped paper nests that hang from trees or house roofs. Like all paper wasps, they chew up pieces of soft wood and masticate this material with saliva to make the nests. The wasp larvae hang upside down in the paper cells and ask for masticated meat food by forming in their mouths a sweet liquid drop that the adults like to sip, so that the adult wasps are actually paid by their babies to bring food to them.

Yellow jackets are almost equally vicious when disturbed, but are smaller, and make their nests for the most part underground, in such places as abandoned gopher holes. One of my sons was attacked by yellow jackets when, at about age seven, he stepped on their nest hole. I had to grab him and rush him off to safety, or he might actually have been killed by too many stings. Hornets, which are bigger and stronger may kill and eat yellow jackets. Hornets and yellow jackets hunt by flying high above the ground, making swift lunges down on insects in the air or on leaves or stems. They use their powerful jaws to kill the insect or worm, break away the hard parts, and chew the soft parts into a paste for their larvae.

Another type of paper wasp is the yellow-legged paper wasp, *Mischocytarrus,* which is yellow and black and has a much smaller abdomen. It is common in buildings, but is much calmer around humans than the hornets and yellow jackets. These wasps often make nests of irregular shape, filling each cell with a wasp grub and meat to feed it. They feed on caterpillars and soft insects. The golden paper wasps, *Polistes,* are common paper wasps. Like *Vespa* wasps, they have long, thick abdomens, but they are much tamer around humans. Their nests often appear only partly constructed, with open cells where they put their larvae and feed them.

Several paper wasp queens will join each other in the spring to form a colony, working together quite amicably. In fact, a queen who has started a colony and has a brood of youngsters growing may learn about other queens nearby and desert her own brood to join them. By sounds, smell, or sight, she gets a message, "Come on over and join us." This sociability is a far cry from the honey bee queen, who kills any rival she meets.

Like bees and ants, wasps learn from experience. The first young that emerge from the nest stay close to home for a few days, watching the adult wasps. By following older wasps, they learn where to find and capture food and how to avoid danger. Their best defense is that they themselves are quite dangerous, which they announce with their bright, striped markings.

All these wasps communicate by buzzing sounds, low when talking with mates, shrill and angry when attacking enemies that disturb them or threaten their nests.

ANTS

Ants appear to be more intelligent than most insects. Instead of living just a few weeks or months, as do most other insects, worker ants and their queens may live for ten years or longer. This longer life span gives them more time to learn from experience. Since they live in very large societies, sometimes of hundreds of thousands or even millions of workers, ants must cooperate to preserve and protect the colony or city. However, even with ants, it is rare to find individuals who display actual intelligence, that is, the ability to make correct choices of conduct to meet an emergency. In most cases it is the colony as a whole, rather than the individual ant, that collectively makes such a wise choice, but there are exceptions.

Though some ants can see fairly well with their comparatively large eyes, other ants have eyes so small that they can hardly see at all; they must depend on their other senses. The terrifying driver ants, for example, are blind, except for their queens; nevertheless, their masses of hungry soldiers and workers often overcome every living thing they meet.

There are many remarkable ants of different kinds. Some communicate best by sight, others by touch or feel, others by hearing. Many use scent devices such as pheronomes that, when left on trails, give orders and suggestions to the ants that find them. By watching ants carefully, one can often find out or make educated guesses as to what they are saying to one another.

Argentine Ants

Argentine ants and the fire ants, both recent immigrants, appear to have particularly complex languages and active intelligences. They

Insects and Their Relatives

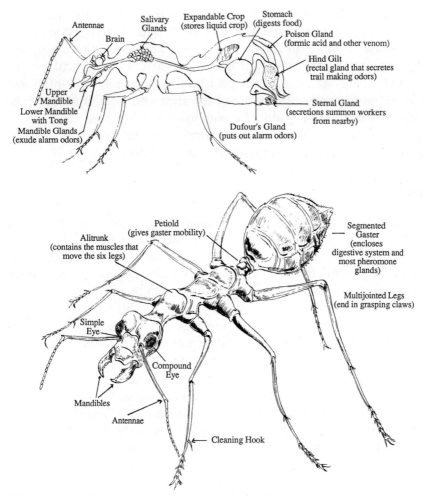

The anatomy of an ant: (above) *glands;* (below) *superstructure.*

entered the United States fairly recently from the south and seem to have spread very rapidly through the southeast, the southwest, and California.

Argentine ants are very small, soft ants, and so it would seem they would be easily overcome in battles with other ants. (As they are great pests in both house and garden, a lot of people wish they could be easily overcome.) In fighting with other ants, they have the advantage of having several queens in each colony, so that having one or two or more queens killed does not doom the colony. But their most important edge of superiority seems to be their exceptional ability to cooperate. Seeming to follow orders, probably optical or aural, several Argentine ants, coming from several directions, suddenly attack a single enemy ant, each seizing a leg until the enemy is helpless and can be cut to pieces. More close observation and study is necessary to understand how this kind of communication is passed from ant to ant. One danger is that the newly arrived Argentine ants may eliminate other ants that are important friends of ours.

Fire Ants

There is, perhaps fortunately, one other kind of ant that seems to be able to use intelligence as well as does the Argentine ant. This is the fire ant, which has spread almost as far in the southern United States as the Argentine. Both helpful and harmful to mankind, it eats some of our crops but destroys insects that are even more dangerous to our crops. It also seems to be the only ant species able to fight and overcome Argentine ants in battle. Fire ants shoot a burning poison out of their hind stings. Seemingly intelligent in several ways, they build their cities under large rocks or under tangles of trunks so that it is very difficult for humans to dig them out.

Argentine ants have invaded widespread areas, but fire ants have been able to stop them by forming battle ranks and shooting poisonous clouds at them from their rear ends, causing the Argentine ants to flee in terror. It would be interesting to find out how the fire ants organize this attack and how the order to fire is given.

Carpenter Ants

These big, black ants are very common around coniferous forests and their edges. They generally make their homes in rotting trees, and using their powerful jaws, build numerous tunnels and rooms in the wood where they can protect their eggs and young. At first when studying carpenter ants, I thought they were not very intelligent, because they seemed to scatter excitedly in all directions when I would open one of their nests in partly rotted wood, fleeing into holes deeper in the wood. Later on, I saw them systematically attacking a mouse that was trying to get at their nest, probably looking for eggs and larvae. They made it so hot for the mouse that it left. After observing this incident, I decided that when I opened the nest, a message had come from the surface to warn the inhabitants that a human was attacking who would be too strong for them to resist. The command went out: "Get the young and eggs and queen to a safer place, quick!" So the first inclination to attack me turned into flight when they realized their danger.

Harvester Ants

Among the more interesting ants are the harvester ants, which live in open plains and valleys. Harvester ants are found mainly in the drier parts of the Western states and in open areas such as deserts or desert grasslands, and occur as far east as western Texas or Kansas. They usually build large mounds out of bits of stone or clay that becomes almost as hard as rock, enabling them to keep water out of their nests underground. They go out seeking for seeds, sometimes as much as a hundred yards from the nest. The seeds are cracked open with their powerful jaws and the kernels kept deep underground in storage rooms. The lives of these ants is simple, for they do not need an elaborate language. Most commands are sent by sight or the sounds of ants moving on the trails, telling the workers where to go for the best crops of seeds. These ants have apparently never learned to plant gardens to produce more seeds. But it has been recorded that in such areas as western Kansas and Oklahoma, where the prairies are more lush, similar ants have learned to raise gardens from seeds to produce more food for their cities. If this is true, it is indeed a forward leap of intelligence.

Leaf-cutting Ants

These ants are among the strangest and most interesting. They are the only ants I know that have underground gardens where they raise almost all their own food. They are often discovered denuding a broad-leaved tree. The giant queen ant usually starts her colony or city in a hole hidden in the ground. Here she lays special, large eggs that are used as fertilizer for her first underground fungus garden,

Harvester ants: an underground city exists beneath the mound.

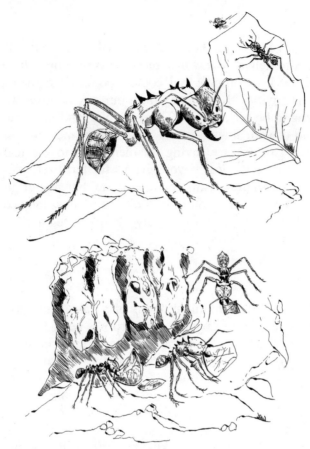

Leaf-cutting ant: (above) *a small guard ant rides on the leaf;* (below) *in the underground nest, ants fertilize the fungus garden.*

which she starts with a special packet of fungus in her body. Then she lays other eggs that hatch workers, which range in size from tiny minims, to average-size workers, to giant soldier ants. The soldier ants are an inch or more long, and have huge jaws for fighting. The worker ants do most of the work of cutting pieces out of leaves, which they carry home and use to fertilize the underground gardens. But just as important are the minims, tiny ants only one-eighth of an inch long who, acting as guards, ride on the cut leaf pieces the workers carry and constantly try to fight off nasty little phorid flies that otherwise attack the workers' heads.

    The minim is plainly saying to the flies, as she rides along with the lumbering workers, "You keep off or I will bite you!" The giant soldier ants also deliver a very firm message as they open their jaws at any animal, bird, or human that comes too close to their precious city. They are saying, "Keep away from our city and our workers or you are going to be bitten *hard!*" They do bite, and very painfully, if they get hold of your finger, as has happened to me!

The fungus gardens are saying, in their way, to the leaf-cutting ants, "In return for the leaves you give us for fertilizer, you can eat our spores." Essentially all ant colonies are exchangers of goods and services, and this is where you must look to find how they communicate. War alone disrupts the exchange and creates a new, hostile communication.

Parasitic Ants

To name even a few of the many kinds of parasitic ants as separate species would be unnecessary for our purposes here. Instead we will look at some of the typical behaviors of the various species. Parasitic ants usually copy the characteristics of the ants that they want to parasitize. The parasitic ant queen, for example, may enter the host city or nest and hop on top of the host queen. Smaller than the host queen, she can rest there and gradually take on her smell. Then, she can say, in effect, to the regular ants, "See, I smell just like your queen," and keep them from attacking her. When she lays eggs of her species, the host ants unknowingly take them up as if they were their own and care for them until they become grubs; the grubs are also fed by the host ants until they pupate and become full-grown parasitic ants, secretly living in the other ant city, accepted because of their smell. Eventually the parasitic queen may kill the real queen and take over the whole city.

Formica Ants

Many of the common red or blackish ants of the *Formica* species are keepers of herds of aphids or other honeydew-producing insects such as scale insects, leafhoppers, and treehoppers. The ants protect these herds as best they can from carnivorous insects such as mantids or lacewings. In return the aphids and scale insects give the ants liquid

Ants: (above) *red* Formica *ants herd scale insects for honeydew;* (below) *worker ant carries a wounded sister ant.*

honeydew from tubes in their rear ends. Besides protecting these insect "cattle" by day, some ants actually induce them, by gentle prodding, to move at night into the ant nests, where they can be better protected. There is thus constant communication between aphids and ants. The ants also try to keep predator insects away from their herds, by rushing at the predators and trying to bite them.

TERMITES

Termites are sometimes called white ants, but they are not really ants at all or even closely related to them; they are related more closely to the cockroach. Wood cellulose is converted into digestible form by protozoans and bacteria that live symbiotically in the termite digestive tract. Termites bite holes through soft or rotting wood and make tunnels and rooms in it to protect themselves from ants and other creatures that might attack them. They form towns or cities of many hundreds or thousands of workers, who do the work of the colony for the kings and queens. The queen is the most important as she lays the eggs that produce the workers.

Something like intelligence is certainly shown among termite cities. Soldiers use their large, hard heads to fight enemies, such as ants, or to plug holes in their tunnels so that enemies cannot get through. Among some species, specialized soldiers called nasutes shoot out nauseous or even poisonous droplets or gasses that injure or kill enemies such as the ants. These termites are saying, "Stay out of our cities!"

While ant workers are all female, termite workers are both male and female. Termite kings are treated with more respect than are ant males, who are usually kicked out of the nests or killed when the queens have been fertilized.

Because they live so well hidden in wood, it is difficult to study termites, but you can make an artificial nest by placing two glass sheets a half inch apart and positioning pieces of wood, through which the termites can burrow, between the pieces of glass. Most of the time, keep black cloths over the glass so that the insects can work in darkness, but take away the cloth at night and use red lights to study how they communicate.

SCORPIONS AND SPIDERS

Scorpions, whip scorpions, and pseudoscorpions all have large pincers, like those of crabs, with which they seize their prey. The scorpion has the additional advantage of a dangerous sting on the end of the tail. Scorpions wave their pincers and lift their tails menacingly as a warning to leave them alone. They can see for only a short distance, but the instant their pincers run into any small creature, they seize it, then whip over the tail and sting it.

Scorpions and their near relatives are most remarkable for the courtship dance of the male and female. The pincers of each pair touch and then seize hold of each other, like a man and woman holding hands at a folk dance. Then they begin a grotesque dance back and forth and around, which may go on for several minutes or longer until the female is ready to mate.

Spiders have several means of communicating with one another, but mainly at mating time, since otherwise they have little to do with each other. They do have one or two ways of communicating a threat to those who would attack them. Both tarantulas and jumping spiders, as well as several other kinds, hold up two or more of their front legs in a threatening fashion and display their sharp fangs, ready to bite. They may even jump a little bit toward their target in a threatening way.

The strangest communicating methods are used by the male spider in his attempt to woo the female. While there are great variations in the wooing of spiders, there are two general types, those done by feel and those done by sight. Most spiders have extremely poor eyesight, and the females can easily mistake a male for an edible insect and attack him quickly. With such spiders, it is necessary for the male to send a tactile message to the female that will convince her that he is a good mate.

Most spiders have organs in their legs that are very sensitive to touch. Other leg organs are sensitive to chemical stimuli in a way that is similar to the senses of taste and smell, but that constitutes a distinctive spider "sense." The male spider approaches the web of the female and sends her a message by pulling on the web. If she is interested, she may pull back. When he gets near her, he touches her with his front feet, and this touch, if done gently and carefully, may convince her in time that he is the one for her.

The tarantula male approaches the female on the ground and begins to hammer at her with his four front legs. These "love taps" finally rouse her to tap back.

Spiders with good eyesight, such as the very interesting jumping spiders, use a combination of arm signals and brilliant colors for mating communication. Jumping spider males, and sometimes the females, are brilliantly colored. Once he has attracted the female by his bright colors, the male begins to wave his arms in fantastic ways and sometimes whirls about in a little dance, in which the female may join him.

Orb Weaver Spiders

The orb weaver spiders have the most beautiful webs in the world; they often glisten like silver when covered with the morning dew. The females are quite big and dangerous, while the males are smaller and easily killed by a female who mistakes one for an insect that has come onto her web. The males have three ways of trying to calm the female and make friends with her. One way is by plucking the web with his front legs in such a way that she gets the message that he is a spider and not an insect. The female has a remarkable sense about vibrations from much practice catching insects, so if the male gives exactly the right vibrations, he is accepted. In more danger is the male who tries to calm the female by his touch. This means he has to come close to her and if he does not do it exactly right, he is a dead spider! A third way is by sound. Somehow, perhaps by rubbing his legs together or with his teeth, he makes soft noises that mollify his prospective mate and allow him to approach her. Such wooing of a female spider by the male is fascinating to watch and to hear, if you have sharp ears.

Wandering Lynx Spider

Among spiders that do not make webs and so cannot use a web to send signals to possible mates, some, like the wandering lynx spider, will actually catch an insect, like a fly or grasshopper, kill it, and wrap it in a silken ball, then place this delicately wrapped parcel in front of a female when she rushes him. This signal stops her from eating him and makes it possible for him to mate with her while she is eating the grasshopper.

Wolf Spiders and Jumping Spiders

Wolf spiders and jumping spiders signal and mollify the female by waving their often brightly colored front legs. Each species has its particular way of leg waving, so that the female will recognize only the one that is for her. Some male wolf spiders have long bristles on their front legs, and these bristles rise erect and stick out of the legs as

Spiders: (above) *courting tarantulas:* (left) *male and* (right) *female;* (below, left) *wolf spider and* (right) *jumping spider, both displaying for a female.*

they move, creating the effect of a leaping forest. The female is fascinated and comes closer!

But it is the male jumping spiders, usually small in size and often colorfully marked on both body and legs, who seem to put on the best displays. Perhaps this is because they have the best eyes of all spiders; larger than usual for spiders, these eyes allow them to see clearly for distances. They can make quite large jumps to grab their prey, so the male has to be sure not to come close to the female until his leg waving convinces her that he is of the same species. Male jumping spiders often use not only their front legs, which are often brilliantly colored, but also their colorful palpi, or front feelers. They may even change the colors of their eyes, making them sparkle brilliantly. The male does not approach the female from only one direction but moves back and forth sideways like a crab, trying to rouse her to mate with him. Such communication between jumping spiders is often very fascinating to watch.

9
Finding the Answers to the Unknown

The secret world of animal language is only beginning to be opened up to man. You can learn the languages of animals about whom almost nothing is now known, or you can, through your exploration and study, vastly increase our present, very insufficient knowledge of those that have been studied in part.

The chief things needed for this study are intense curiosity, time, and a willingness to investigate carefully and observe the animal you are studying until you unravel the secrets of its language and life. To do this correctly, you should follow the steps suggested in this book and learn how to use the scientific method, which means simply applying systematic common sense, looking at things from all sides, and remembering never to jump to conclusions but to come to them only after long and careful testing and research.

It is wise to remember that animals are different from human beings, often far different in their reasoning powers and in their means of communication. Some of them observe tiny details that we would naturally miss. Others have a keen sense of smell and so read messages brought by the breeze that have no meaning to us at all. With nearly all animals, it is extremely difficult to discover and understand all their secrets of communication.

Many animals live in environments almost beyond our ability to reach and study, such as the mole and gopher in their underground homes, or the bats who fly through the hours of darkness. But we would be poor naturalists and poor human beings if we allowed such difficulties to discourage us. Naturalists have found that, by using a dark red flashlight beam at night, they can study the lives of many nocturnal animals without disturbing them. You, too, might make similar discoveries.

Perhaps one of the best ways to study animal languages is to go into the places where they live and become so intimately acquainted with them that you can make friends and learn to talk to them. A lady in England turned her house in the country into a house for birds. The wild birds came and went in it as free as the air and fed from her hands. She learned many intimate secrets of their lives. From my own animal friends, I have learned much of what has gone into this book.

When you talk to animals, it is best to use a high voice or a very soft voice, since they associate such sounds with their babies.

Sometimes you can talk by mechanical means, such as through bird whistles, crow calls, duck calls, moose calls, and so forth, which can be purchased at sporting goods stores and museum stores. You need to practice a great deal with such calls in order to get the exact tones of a wild animal or bird.

Practicing with just your voice, you may be able to imitate the bark of a fox, the squall of a wildcat, the chattering of a red squirrel, or the whistling of a bird. By sitting perfectly still when you do it, you may bring these animals near you. If you put out food they like and sit quietly, they may become tame enough in time to come close to you, especially if they sense your friendship and no harm comes to them.

In the Panama jungle, I often watched wild animals by hiding in dense clumps of ferns on the tops of small cliffs, from which I could look down into a creek canyon below. Since I was so high, my human scent was carried away by the breeze. To make myself doubly inconspicuous, I wore clothes colored like the ferns and rubbed these clothes with strong-smelling weeds so as to disguise the human scent.

Sometimes better results can be had by building carefully hidden blinds. The accompanying illustration shows three types of blinds. Notice that inside the blind there is a comfortable place for a man to sit or to lie, and that there are peep holes through which he can watch on all sides. It is essential to get comfortable inside such a blind and then to be as still as possible. Rub strong-smelling plants on your clothes and body to disguise the human scent. Clothes boiled in water with such plants work even better.

A spring in the desert or other dry country is a marvelous place to

Various blinds for watching wildlife.

watch for wildlife, especially in the early morning or evening, as many animals come then to drink. A location where there are numerous animal burrows is another good place to build a blind.

You may need to be quiet for a long time. The older a blind gets, the more used to it the animals become, until finally they will pay no attention to it at all.

Always keep a notebook handy when you are observing things in the wild. Write down the date, time, place, and appearance of your surroundings and the state of the weather. The last is very important, as it often has a good deal to do with the feelings of the animals and what they say to each other. Then, as animals and birds appear, note their various sounds and actions, trying to catch the emotions they express. Be very careful not to assume that the animals think the way you do.

Sometimes, of course, it is not possible to keep records in the field that are as full as you would like, because you must watch happenings closely or because the sound of the pencil is disturbing. In this case, write down everything you can remember as soon as you get home or back to camp. All such records can be very useful to science.

It is important for the true scientist to do a great deal of research into what other people have written about animals. It is also important to be cautious about believing everything they say. For example, the famous naturalist Jean-Henri Fabre once stated that spiders are completely instinctive in all their actions and can learn nothing from experience. Later evidence has proven him wrong. Be careful, then, about believing all I have written in this book, because I, too, could be wrong. Be careful: be scientific!

Index

Alligators, 160
Amphibians, 164–66
  frogs, 165–66
  newts, 164–65
  salamanders, 164–65
  toads, 165–66
Animal communications chart, 8
Animal-human communication, 12
Animal language, generally
  attack, effect of signaling, 19
  complex forms, where found, 18–19
  factors affecting, 6, 190
  imitating sounds, 191
  instinctive nature of, 5
  mating season and, 6
  miscellaneous, 12–17
  recognizing animal talk, 2
  scientific approach to, 7–24
  social communication, forms of, 18–24
  territorial influences, 6
  weather, effect, 6
Animal movements, meanings, 2
Animal-plant communication, 13
Antelopes, longhorn, 110–13
Ants, 178–86
  Argentine ants, 178–80
  carpenter ants, 180–81
  fire ants, 180
  formica ants, 184–85
  harvester ants, 181
  leaf cutting ants, 181–82
  parasitic ants, 184
Armadillos, 130
Attacking
  actual attack by mobbing, 11–12
  mob action, signalling, 11
  similar but different species, 11
  within the species, 11

Badgers, 59–60
Bats, 129

Bear family, 91–99
  black bears, 91–95
  grizzlies, 95–96
  language of movement, 92
  raccoons, 96–99
  sounds made, meanings of 94
Beavers, 122–24
  danger, warnings of 123
  intelligence, 122
  work, cooperative nature of, 124
Bees, 170–76
  artificial hive, constructing, 22
  complicated language of, 171
  food, use in communication, 171–73
Bighorn sheep, 113
Birds
  bitterns, 143
  bobwhites, 150
  canaries, 54–55
  crows, 155–57
  doves, 149–50
  ducks, 141–42
  eagle, 145
  events that influence language, 133–38
  flickers, 152
  flycatchers, 154
  geese, 141–42
  grebes, 143–44
  gulls, 138
  hawks, 145–48
  herons, 143
  hummingbirds, 144–45
  jays, 155–57
  loons, 143–44
  nighthawks, 148
  owls, 150–52
  parakeets, 54–55
  pecking birds, 154–58
  pigeons, 149–50
  roadrunners, 152
  sea birds, 138

Index

Birds (*continued*)
   shore birds, 140–41
   swans, 141–42
   swifts, 148–49
   vultures, 145
   wild, 131–58
   woodpeckers, 152–53
Bitterns, 143
Black bears, 91–95
   language of movement, 92
   mating behavior, 95
   sounds made, meanings, 94
Bobwhites, 150

Camouflage, use of, 16–17
   mimicry and, 15–16
Canaries, 54–55
Carnivores
   escaping, 18
   signaling danger, 18
Cat family, 31–34, 68–75
   curiosity of, 34
   fighting stance, 32–33
   lynx, 71–72
   mountain lions, 72–75
   self-grooming, reasons, 34
   ways of communicating, 33–34
   wildcats, 69–71
Cattle
   bulls, warning signs prior to attack, 39–41
   cows, 40
   curiosity of, 40
Chickens, 46–50
   calls given by, 47
   mating calls and actions, 50
   pecking order, 47–49
   protecting eggs, 49
   protecting chicks, 50
   sounds made by, 46 *et seq.*
Coyotes, 79–82
   packs, 79–81
   scent signals, 82
Crows, 155–57
Curlews, 140–41

Deer family
   banding, effect, 100
   elk, 106–09
   members of, 100–10
   moose, 109–10
   mule deer, 104–06
   white-tail deer, 101–04
Dogs, 25–30
   fighting, signs indicating, 27–28
   jackal blood in, effect, 26–27

Dogs (*continued*)
   language of, 25–30
   mating habits, 29
   smelling and hearing capabilities, 27
   understanding, factors affecting, 26
   variety of ways of communicating, 27–28,30
   wolves, reactions like, 30
Domestic animals, languages of, 25–55
   See also specific subject headings
Doves, 149–50
Ducks, 53–54, 141–42
Eagles, 145
   methods of catching food as affecting language, 145
Elk, 106–109
   bugling cry, 107
   harem system, 106–07
   rump hair, raising, 108
Emotional language of wild animals, understanding, 5
Enemies, escaping from, 17–18

Fish, 166–68
   schools, communication by scales, 166–67
Fixed action display, 9–11
   male trying to attract mate, 9–10
   territory, guarding, 10
Flickers, 152
Flycatchers, 154
Freezing. *See* Indecision
Frogs, 165–66

Geese, 141–42
Gray fox, 86–87
Grebes, 143–44
Grizzlies, 95–96
Gulls, 138

Harbor seal, 88
Hares, 114–16
   rabbits distinguished, 114
Hawks
   bird hawks, 147
   falcons, 147–48
   harrying hawks, 147
   method of catching food as affecting language, 145
   soaring hawks, 145
Herons, 143
Horses, 34–39
   affection toward masters, 34–35
   fear, manifesting, 37
   intelligence, 34
   wild stallions gathering mares, 35–36
Hummingbirds, 144–45

Index

Indecision
  effects, 16–17
  freezing, 16–17
Insectivores, 129–30
  moles, 129–30
  shrews, 130
Insects, 169–89
  ants. *See* Ants
  bees. *See* Bees
  scorpions, 186–87
  spiders. *See* Spiders
  termites, 185–86
  wasps. *See* Wasps

Jays, 155–57

Killdeers, actions, 3–5, 140
Kit foxes, 87

Leopard seal, 88
Listening, need for, 2
Lizards, 161–62
Long distance signals, 12
Loons, 143–44
Lynx, 71–72

Mammals
  characteristics of, 100
  vegetarian. *See* Vegetarian mammals
Meat-eating animals, 56–99
  See also specific subjects
Mice, 116–19
  grasshopper mice, 119
  white-footed mice, 116–19
Migration, use of communication skills during, 14
Mimicry, use of, 15–17
Minks, 59
Moles, 129–130
  size, 109
  wallows, 110
Mountain lions, 72–75
Mule deer, 104–06
  calls, 105
  scent glands, use in communicating, 106

Newts, 164–65
Nighthawks, 148–49
  poorwills, 148–49
  whipporwills, 148–49

Otters, 60–63
Owls, 150–52

Parakeets, 54–55
Peccaries, 114

Pigeons, 149–50
Pigs, wild native, 114
Play, language of, 16
Poorwills, 148–49
Porcupines, 128–29
Prairie dogs, black tailed, 124
Pronghorn antelopes, 110–13
  hair, use in signaling, 111
  smell glands, use, 111
  sounds made by, 113
  speed, 110–11

Quail, 150

Rabbits, 44–46
  hares, differences between, 114
  male/female behavior, 45
  nest making, 45
Raccoons, 96–99
Rats, 119–122
  kangaroo rats, 122
  pack rats, 121
  Norway rats, 119
  wood rats, 119–22
Red foxes, 83–86
  intelligence of, 85
  scent posts, 86
Reptiles, 159–64
  alligators, 160
  lizards, 161–62
  rattlesnakes, 162
  snakes, 162–64
  turtles, 159–60
Roadrunners, 152
Rodents, generally, 116–30
  See also specific subjects

Salamanders, 164–65
Scorpions, 186–87
Scientific approach to animal language, 7–24
Sea birds, 138
Sea lions, 88–91
Sheep, 41–44
  "baa", meanings expressed by, 44
  intelligence, 41
  rams, fighting nature of, 44
  wild, 113
Shore birds, 140–41
  curlews, 140–41
  killdeers, 140
Shrews, 130
Skunks, 63–68
  smell, signals given by, 64
Snakes, 162–64

Index

Social Communication
  ants, slave-making among, 20–21
  city and colony structures, constructing, 22–23
  feeding time, signaling, 20
  foraging, cooperative nature of, 24
  migrating animals, 23–24
  nest building, 21
  pheromones, use by ants, 23
  social behavior, teaching to the young, 20
  social defenses, examples, 19
  trail making, message conveyed by, 19
Spiders, 187–89
  jumping spiders, 188–89
  orb weavers, 188
  wandering lynx spiders, 188
  wolf spiders, 188–89
Squirrels, 124–29
  black tailed prairie dog, 124
  gray squirrels, 125
  ground squirrels, 127
  porcupines, 128–29
  red squirrels, 126
  woodchucks, 128
Swans, 141–42
Swifts, 148–49

Termites, 185–86
Toads, 165–66
Turkeys, 50–53
  danger, reaction to, 52
  fights, 51
  mating behavior, 52
  pecking order, 51
  tame and wild turkeys compared, 51–52
Turtles, 159–60

Understanding animals, effect, 1

Vegetarian mammals, 100–30
  See also specific subject headings
  characteristics of, identifying, 100
Vultures, 145

Wasps, 176–178
  social wasps, 176
  solitary wasps, 176
  sphecid wasps, 176
Weasels, 56–68
  badgers, 59–60
  minks, 59
  otters, 60–63
  skunks, 62
  true weasels, 57–58
Whipporwills, 148–49
White tailed deer, 101–04
  antlers, 102–03
  calls, 104
  habitat, 101
  mating, 103
  wallows, 102
Wild birds, 131–58
  alarm and distress notes, 132
  feathers, use, 132
  simple nature of language of, 131–32
Wild cats, 69–71
Wild dogs, 76–79
  complexity of language of, 77–79
  coyotes, 79–82
  gray foxes, 86–87
  kit foxes, 87
  pack hunters, 75–76
  red foxes, 83–86
  wolves, 76–79
Wild birds, 131–58
Wild pigeons, 149–50
Wild sheep, 113
Wolves, 76–79
  complexity of language, 77–79
  organization of pack, 76
Woodchucks, 128
Woodpeckers, 152–53

Young of the species, teaching survival skills to, 13–14